Epidemiological
Research Methods

Epidemiological Research Methods

Don McNeil
Macquarie University, Sydney, Australia

JOHN WILEY & SONS

New York • Chichester • Brisbane • Toronto • Singapore

Other Wiley Editorial Offices

John Wiley & Sons, Inc., 605 Third Avenue,
New York, NY 10158-0012, USA

Jacaranda Wiley Ltd, 33 Park Road, Milton,
Queensland 4064, Australia

John Wiley & Sons (Canada) Ltd, 22 Worcester Road,
Rexdale, Ontario M9W 1L1, Canada

John Wiley & Sons (SEA) Pte Ltd, 37 Jalan Pemimpin #05-04,
Block B, Union Industrial Building, Singapore 2057

British Library Cataloguing in Publication Data

A catalogue record for this book is available from the British Library

ISBN 0 471 96195 7; 0 471 96196 5 (pbk)

Produced from camera-ready copy supplied by the author.
Printed and bound in Great Britain by Biddles Ltd, Guildford and King's Lynn.
This book is printed on acid-free paper responsibly manufactured from sustainable forestation,
for which at least two trees are planted for each one used for paper production.

Contents

PREFACE   ix

1. **EPIDEMIOLOGICAL RESEARCH**   1
    1: Introduction   1
    2: Measurement   3
    3: Study Types   6
    4: The Credibility of a Study   11
    5: Sampling Variability   15
    6: Statistical versus Clinical Significance   23
       Summary   25
       Exercises   26
       References   28

2. **STATISTICAL METHODS I**   31
    1: Introduction   31
    2: Two-by-two Tables   32
    3: Multiple Outcomes   39
    4: Continuous Outcomes   41
    5: Paired Data   46
       Summary   54
       Exercises   55
       References   58

3. **STATISTICAL METHODS II**   61
    1: Introduction   61
    2: General Contingency Tables   62
    3: One-way Analysis of Variance   66
    4: Two-way Analysis of Variance   71
    5: Simple Regression   77
    6: One-way Anova by Regression   82

Summary 89
Exercises 90
References 92

4. MANTEL–HAENZEL METHODS **95**
1: Introduction 95
2: Confounding in 2-by-2 Tables 99
3: Combining Odds Ratios 104
4: The Relative Risk 110
5: Multiple Risk Factors 115
Summary 120
Exercises 121
References 123

5. LOGISTIC REGRESSION **125**
1: Introduction 125
2: Modelling Confounding 130
3: Modelling Stratified Data 140
4: Multiple Risk Factors 146
Summary 154
Exercises 155
References 157

6. LOGISTIC REGRESSION II **159**
1: Introduction 159
2: Case-by-Case Data 160
3: Modelling Risks 164
4: Poisson Regression 170
5: Modelling Outcome Severity 181
6: Survival Data 185
Summary 189
Exercises 190
References 192

7.	SURVIVAL ANALYSIS	195
	1: Introduction	195
	2: The Survival Curve	197
	3: Comparing Survival Curves	200
	4: The Proportional Hazards Model	208
	5: Hazard versus Survival	212
	6: Modelling Covariates	216
	7: Other Models	221
	Summary	227
	Exercises	228
	References	232
8.	MATCHING	235
	1: Introduction	235
	2: Matched Pairs	238
	3: Logistic Modelling	245
	4: Before–After Studies	251
	5: 1:M Matched Case-Control Studies	254
	6: Pros and Cons of Matching	257
	Summary	261
	Exercises	262
	References	264
9.	SAMPLE SIZE	265
	1: Introduction	265
	2: Precision of an Estimate	266
	3: Power of a Study	271
	4: Clinical Trials	279
	5: Meta-analysis	282
	Summary	288
	Exercises	289
	References	289
	APPENDIX	292
	INDEX	297

Preface

My interest in epidemiology started in 1984, when the Institute for Advanced Research in Cancer (IARC) ran a short course on cancer epidemiology in the Department of Public Health at the University of Sydney. The keynote speakers for this course were Nick Day and Alec Walker, and the material they covered in their lectures opened up a new world to me.

As a statistician I had been involved in collaborative work with medical researchers, including consulting for various pharmaceutical companies during my years at Princeton University in the US, and then after coming to Sydney acting as consulting statistician for the Australian - New Zealand Breast Cancer Trials Group. Learning about clinical trials from this group was exciting enough, and I was already beginning to incorporate the material into my lectures at Macquarie, but the IARC course convinced me that what my university needed was a fully-fledged programme in biostatistics and epidemiology.

This book has resulted from my efforts to develop a programme in epidemiology within a Department of Statistics. It is thus designed for undergraduate students in Statistics, and the material covered is taught at both the final-year undergraduate level and at the Masters level.

The book is also designed for medical scientists. In 1987 as part of an international aid program I started teaching at Prince of Songkla University (PSU) in Thailand, and four years later Dr Virasak (one of the participants in the 1984 IARC course) established an international MSc program in Epidemiology in the Medical School at PSU. The first draft of *Epidemiological Research Methods* arose as the handouts for my lectures for these students (and for similar students at Mahidol University in Bangkok), comprising medical graduates from various countries in the region, and data collected by these students for their dissertations are used as illustrations of some of the methods. Some of the material also originated from lecture notes for training courses given at the NHMRC Clinical Trials Centre at the University of Sydney.

Observation and experimentation are the two cornerstones of scientific research. Traditional epidemiological methods, such as those described in the books by Breslow and Day and by Kleinbaum, Kupper and Morgenstern, have tended to focus on observational studies, with other books devoted to experimental studies (such as Pocock's book on clinical trials). Yet the statistical methods needed for the two types of investigation are much the same, and there seems no compelling reason for separating them. Indeed there are excellent reasons for studying them together. (In doing so you will discover, for example, that the logrank test used to compare the survival prospects of patients in experimental trials is a special case of the celebrated Mantel–Haenszel test used to assess the risk associated with an observational exposure in the presence of a confounding variable.)

This book is perhaps ambitious in that it attempts to cover in reasonable depth the concepts of statistical modelling of epidemiological data, including logistic and Poisson regression for both matched and unmatched designs as well as survival analysis, with minimal statistical prerequisites. The rationale for this attempt is that while a deductive approach based on theory is desirable for a student of Mathematics, an inductive approach based on examples and data is preferable for almost everyone else. And given that fast computers with user-friendly software are now available to all students, the data-based approach is feasible.

Many data sets are used to illustrate the methods in this book. A few are hypothetical, but the rest are chosen from real-life scientific studies. Many have been analysed before. With just one exception (where the full data set is given in the 1980 book on case-control studies by Breslow and Day) all data are listed. A readily available source of data from published scientific studies is the book *A Handbook of Small Data Sets* by Hand, Daly, Lunn, McConway and Ostrowski, and many of the medical examples from this book are used as exercises.

I would like to thank some of my friends and colleagues for their encouragement with this book; in particular Geoffrey Berry, Amornrath Podhipak and Roslyn Poulos for their useful comments on the first draft, Virasak Chongsuvivatwong, Chamnein Chounpradub, and John Simes for providing study havens where I could write and think away from the distractions of my home department, Elaine Beller, Abie Ekangaki, Alan Geater, Val Gebski, Thomas Lumley, Mike Jones, David Kleinbaum, Oberon Leung, Dan Lunn, Mary Lunn, Robyn Richards, David Signorini and Marvin Zelen for their interest and help, and not least my graduate students Halimah Awang, Jiraporn Chompikul, Rohana Jani, CK Leung, Kehui Luo and Roslyn Poulos.

Don McNeil

November, 1995

1

EPIDEMIOLOGICAL RESEARCH

1: Introduction

This book is an attempt to explain what epidemiological research is about. Epidemiology is the branch of Medicine concerned with understanding the factors that cause, prevent, and reduce diseases by studying associations between disease outcomes and their suspected determinants in human populations. It thus requires measuring and comparing rates of disease in populations in which different persons or groups have different exposures to possible determinants.

In fact epidemiological methods have wider applicability: the populations studied need not be human, and the outcomes need not be restricted to diseases. In this book, however, we shall concentrate mostly on medical applications; these have been the catalyst for the rapid expansion of epidemiological research methods in the last decade. We will show how these methods apply to both observational and experimental studies.

One of the first persons to undertake epidemiological research was Louis (1836), who investigated the effect of the ancient and entrenched medical practice of blood-letting on patients with pneumonia and found evidence that delaying the treatment reduced the mortality rate. Largely as a result of this study blood-letting lost favour. Louis (1834) had earlier described many of the basic principles underlying experimental research in epidemiology. However, it was not until the middle of the next century that a major clinical trial was undertaken, when the British Medical Research Council investigated the effect of streptomycin treatment for tuberculosis, the methods being described in detail by Hill (1951). A further landmark was the observational study by Doll and Hill (1964) demonstrating an association between the level of smoking and the mortality rate among British doctors.

Another pioneer was Snow (1855) who found that the cholera death rate in London in 1854 was greater by a factor of 5 among residents drinking water from a particular supplier, thus identifying contaminated water as a risk factor for this disease.

Epidemiology involves taking measurements from subjects and making inferences about relevant characteristics of a wider population typifying the subjects. Since Statistics is the science that is concerned with making inferences about population parameters using sampled measurements, statistical methods provide the basic tools for epidemiological research. Good epidemiological research thus requires both an understanding of Statistics, which provides a means for assessing the probabilities associated with apparent associations between determinants and disease outcomes, and a knowledge of Medicine, enabling statistically probable associations to be interpreted and placed in a biologically causal framework.

This book attempts to be comprehensible and useful to medical researchers and interesting to statisticians. It takes advantage of the fact that modern user-friendly computer packages have made it much easier to analyse data from epidemiological studies, enabling researchers to focus on the most important issues without getting bogged down in statistical formulas or numerical calculations. It also attempts to integrate the statistical methods with the medical interpretations.

In the present chapter we introduce the basic concepts in epidemiological research. These include deciding what to measure and defining the research objectives, designing a research study, controlling factors that may reduce the credibility of an investigation, and understanding the role of statistical variance and the difference between what is statistically detectable and what is of practical importance. These concepts set the scene for the chapters that follow.

Chapters 2 and 3 cover some of the most basic statistical techniques for analysing data. These methods are well known and may be found in modern texts on medical statistics such as the introductory text by Altman (1991) or the more comprehensive work of Armitage and Berry (1994). They deal with the quantification and interpretation of an association between a determinant and an outcome variable in the absence of other variables that may affect the association. Chapter 2 is concerned with the simplest situation in which the determinant variable is dichotomous, while Chapter 3 extends these methods to more general types of determinants.

In epidemiology the outcome of interest is usually disease occurrence, which is dichotomous rather than continuously varying. Chapters 4, 5 and 6 are concerned with the analysis of dichotomous outcome data. In Chapter 4 the traditional Mantel–Haenszel methods for analysing such data are covered. Logistic regression, which provides an alternative and complementary approach to the analysis of dichotomous outcome data, is introduced in Chapter 5. Further aspects of logistic regression including its application to individual case (rather than grouped) data and the Poisson regression model are discussed in Chapter 6.

The remaining three chapters deal with special topics of particular importance in epidemiology. Survival analysis, a method commonly used in clinical trials, is the subject of Chapter 7. Chapter 8 deals with matched study designs. Finally Chapter 9 focuses on the question of determining the sample size of a study with a view to achieving specified precision or statistical power, and gives a brief introduction to meta-analysis, a method used to combine results from several studies investigating a research question.

2: Measurement

In epidemiology, the individual subject is typically the observational unit from which data are collected. Measurements are of two kinds: (a) outcomes or responses, and (b) possible determinants. An outcome is a measure of a subject's disease status at a particular time, whereas a determinant is a cause of the outcome through its action at an earlier time.

A typical outcome is a heart attack, and a possible determinant for this outcome is high blood pressure. Another example of an outcome is lung cancer, for which smoking is known to be a determinant.

Both kinds of data may be either categorical or measured on a continuous scale. It is very common for epidemiological outcomes to be dichotomous, corresponding to a disease being either present or absent. This is convenient from the medical diagnostic point of view: if a subject is diagnosed as having the disease in question, a certain treatment may be justified, but not otherwise. On the other hand important information could be lost by grouping responses into just two categories. For infants with diarrhoea, it is useful to have three categories of disease status (mild, moderate and severe) depending on the extent of the disease. For patients with AIDS the CD4 count (typically ranging from 0 to 400) is one of the most useful measures of health status. The duration of survival is an outcome variable of major interest in many clinical studies.

Determinants take various forms. They include genetic factors affecting predisposition to disease, such as haemophilia which may increase the risk of HIV infection through a contaminated blood transfusion, and demographic factors, notably age and gender. They include environmental and occupational exposures such as contaminated water, asbestos dust, and excess salt and fat in the diet. Behavioural determinants include tobacco and alcohol consumption, exposure to the sun, unsafe sexual practice, and drug addiction. A determinant could also have a positive effect. Thus preventative measures for risk reduction, such as vaccines for combating infectious diseases, screening for cancer, and health promotion campaigns, all fall into the general definition of a determinant. Going further, a determinant could be a treatment aimed at curing or alleviating a disease or preventing further deterioration in a person's health. In this general definition, determinants include medications such as aspirin, hypertensive and hormonal drugs, and treatments such as radiotherapy for cancer and surgery for heart disease.

The distinction between a determinant and a response is not always clear cut. Inadequate prenatal care (measured by the number of visits to a clinic) during the first trimester of pregnancy has been shown to increase the risk of perinatal mortality. However, reduced prenatal care during the third trimester of pregnancy has been associated with a *reduction* in perinatal mortality. The reason for this apparent anomaly is that any pregnancy complication is likely to result in additional prenatal care, so that prenatal care in the first trimester is a determinant, but prenatal care during the third trimester is an outcome. The term *intervening variable* is used to describe a covariate that is on the causal path between a determinant and an outcome.

Individuals can have more than one disease outcome, and outcomes may have multiple causes, and the paths linking determinants to disease outcomes may be intertwined in complex ways. A determinant that prevents or alleviates one disease may cause another. As a result, epidemiological research can present difficult challenges.

A basic objective in epidemiological research is measuring the level of disease in a population. If the outcome is simply disease presence or absence, then the *prevalence* of the disease is of interest: this is simply the proportion of individuals affected with the disease. For outcomes having a continuous range of variation, some statistical summary such as the mean or median may be used to represent the status of the population. Typical examples include (a) the mean birth weight of newborn babies at hospitals in an urban population, and (b) the median survival time for patients in a heart transplant program.

A problem with measuring disease prevalence in a population is that it is likely to reflect what has happened in the past rather than the current situation. In HIV research, the extent to which new cases are occurring is of primary interest, for this measures the effectiveness of preventative programs. Thus determining the *incidence* of a disease, defined as the proportion of new cases of a disease occurring per unit time among persons free of the disease in a population, is another important research objective. Similarly, the median survival time of patients *currently* getting heart transplants may be of primary interest.

While measuring levels of disease status in a population is a basic objective, the most important research objectives in epidemiology involve measuring associations between possible determinants and disease outcomes. For example, it is of interest to know whether taking oral contraceptives affects a woman's risk of developing breast cancer, and many studies have addressed this question. Again, a researcher may be interested in knowing whether there is an association between increased diastolic blood pressure and diabetic illness as measured by the glomerular filtration rate. Where an association has been definitely established, it is of interest to quantify it. For example, what is the relative risk of developing lung cancer for a heavy smoker compared to a non-smoker?

If the outcome of interest is measured on a continuous scale, its association with a risk factor may be expressed as a correlation coefficient, a measure that ranges from −1 (perfect negative association) to 1 (perfect positive association), with a correlation coefficient of 0 indicating no association. If the outcome is a categorical variable, it is more convenient to express its association with a possible determinant in terms of the risk, or probability, that the outcome belongs to a specified category.

If R_1 and R_2 are the probabilities (or 'risks') of a particular outcome for two individuals with two specified levels of a determinant differing by a unit amount ($D_1 = 0$ and $D_2 = 1$, say), the relative risk is defined simply as the ratio R_1/R_2. This is a useful measure of association between a determinant and a disease outcome. A related measure, more useful in many epidemiological studies, is the *odds ratio*. Since the odds associated with a probability p is $p/(1-p)$, the odds ratio is defined as the ratio O_1/O_2, where $O_i = R_i /(1-R_i)$. Another parameter which is widely used, particularly in public health policy, is the *risk difference* R_1-R_2; this provides a measure of the risk attributable to one factor over another. These measures are elaborated and methods for their analysis are considered in detail in the chapters to follow.

Questions of Interest

Where an association is suspected but has not yet been established, the research question is framed in terms of a *null hypothesis*, which states that there is no association between a specified determinant and a particular disease outcome in the population of interest. On the question of a possible association between oral contraceptive use and breast cancer incidence, the null hypothesis would state that there is no association between these two variables in the target population.

The method for testing a null hypothesis is essentially statistical, and is addressed later in this chapter. It is important to realise that a study may not provide a conclusive answer to the question, due to the limited size of the sample. Consequently the result of a study is not certain, but *probable*.

If there are several possible determinants for a disease outcome, an investigator may wish to examine combinations of these determinants, some of which may be associated with the disease. In this case there are several null hypotheses to be tested. These hypotheses need to be ranked in order of importance before the study is undertaken, for given that the conclusion from each hypothesis has a probability of error, the more hypotheses tested, the greater the probability of making at least one erroneous conclusion. There is a limit to the amount of information that can be obtained from any study. Further studies may be needed to settle the lower ranking questions.

3: Study Types

Research questions of interest are investigated by undertaking a study involving the analysis of data measured from a sample of subjects selected from a target population. A study may be *experimental*, in which case the investigator has some control over a determinant, or *observational*, where the investigator has no such control. Studies are further classified according to the method of selection of subjects and the extent of follow-up over time.

Observational Studies

If the research objective is to determine characteristics of the target population, such as the prevalence of a disease or the average of some measure of health status, a *descriptive* study could be undertaken. This involves choosing a representative sample of subjects from the target population, measuring their outcomes, and possibly classifying their prevalence by demographic factors such as age and gender.

Now suppose you wish to go further and investigate a possible association between a suspected determinant and an outcome of interest in the target population. The simplest type of study which allows such an association to be measured is a *cross-sectional survey*, in which subjects are sampled from a target population and classified with respect to both outcome and exposure to determinants of interest. In fact Snow's (1855) investigation of risk factors for cholera was a cross-sectional survey; here the study sample comprised all the residents of a particular area of London where a cholera epidemic had occurred during July and August 1854, and the subjects were classified according to death from cholera (the outcome) and their source of drinking water – Company A (Southwark) or Company B (Lambeth).

A cross-sectional study design is reasonable to use if the determinants of interest do not change very much with time for a given subject. If a determinant can change with time, a cross-sectional survey can give misleading results, as the following hypothetical illustration shows. Suppose you wish to investigate a possible association between heart disease and jogging. Based on a cross-sectional survey of corporate executives aged 50–60 in a particular city you find that only 2% of the joggers have heart disease compared with 10% of the non-joggers. Would you be justified in concluding that jogging reduces the risk of heart disease?

Actually such a conclusion would not necessarily be justified in this case, because it is possible that persons who have developed heart disease have given up jogging due to their reduced fitness. In fact it could be that jogging *increases* the risk of heart disease.

How could the issue be resolved? One approach would involve comparing the incidences of heart disease among joggers and non-joggers. This involves introducing the time (*longitudinal*) element, leading to another type of epidemiological study, the *cohort study*.

A cohort study is an epidemiological investigation in which subjects are selected before they have experienced the outcome of interest, and their exposures to possible determinants of interest are then recorded, together with their subsequent disease outcome. A cohort study thus compares disease incidence in populations classified by the determinant of interest. Breslow and Day (1987) have given a detailed account of the use of cohort studies in cancer research.

Imagine how a cohort study could be designed to investigate jogging as a possible risk factor for heart disease. First, select some joggers and some non-joggers, all free of heart disease, and record the numbers in each group who develop heart disease during a specified follow-up period (5 years, say). Then compare the incidences of heart disease in the two groups.

Will this method identify an association, if one exists, between jogging and heart disease? It will if the two groups are similar in every respect except for their exposure to the determinant of interest (jogging in this case). For then any difference in outcome must be due to the risk factor alone. However, in a cohort study there is no guarantee that the groups are comparable. Suppose, unbeknown to you, most of the joggers are non-smokers and most of the non-joggers are smokers. Given that smoking is a known risk factor for heart disease, any association found in the cohort study between jogging and heart disease could be due entirely to the smoking factor.

Cohort studies often involve the monitoring of a population over an extended period of time, and consequently they are useful for investigating multiple determinants and outcomes. A classic cohort study is the *Framingham Study* in which residents of the town of

Framingham in Massachusetts have been continuously monitored since 1948 with respect to many different risk factors and disease outcomes. This study, described by Dawber (1980), has stimulated a lot of epidemiological research.

If the data collection is done prospectively, a cohort study could be an expensive and time-consuming exercise, particularly if the outcome does not have a high incidence rate. For the hypothetical study investigating the association between jogging and heart disease, it is likely that thousands of subjects would need to be monitored for many years to get sufficiently many events to obtain a conclusive result.

There is a third type of study, called a *case-control study*, which overcomes some of these difficulties. It differs from a cohort study in only one way: instead of selecting subjects for inclusion in the study according to their exposure to the possible determinant and then observing their outcomes, you select subjects according to their outcome status and then measure their exposure to the determinant. In the jogging example, this would involve first selecting a group of company executives aged 50–60 who have had a heart attack in the last 5 years (the *cases*), then selecting another group of 50–60 year old executives, similar to the first in all respects except that they have not had a heart attack in the last five years (the *controls*), and then measuring the risk factors in the two groups. If, for example, it turned out that 10% of the cases were smokers compared with only 5% of those in the control group, this would indicate an association between the risk factor and the disease outcome.

Case-control studies are described in general terms by Schlesselman (1982) and with particular application to cancer research by Breslow and Day (1980).

Cohort studies and case control studies both involve selection. In a cohort study you first select a group of disease-free subjects exposed to the determinant of interest, then you select a comparable group of disease-free subjects not exposed to the determinant, and finally compare the subsequent disease outcomes in the two groups. In a case-control study you first select a group of subjects with the disease, then you select a comparable group of disease-free subjects, and finally compare the prior exposure to the risk factor in the two groups.

If the disease outcome is rare, a cohort study is uneconomical, because a large number of subjects will be needed to obtain sufficiently many outcomes to obtain a conclusive result. In this situation a case-control study is more efficient because one of the two groups being compared consists only of subjects with the disease outcome.

As an illustration, suppose you wish to investigate a possible association between Down's syndrome (a rare birth defect) and maternal smoking. Suppose that the proportion of outcomes is 1% and

25% of the mothers smoke in the target population. For a cohort study design 5000 mothers would need to be studied to expect 50 outcomes. However, it may be shown that a case-control study involving just 100 cases and 100 controls will give approximately the same efficiency.

By the same token a case-control study design is inefficient if the risk factor of interest is rare and the outcome is relatively common. If you wished to investigate diabetes in the mother (a rare occurrence) as a possible determinant of abnormally large birth weight (a less rare occurrence), a cohort study would be a more efficient study design.

Perhaps the most striking example of an effective case-control study is that which identified thalidomide, a drug used by pregnant women, as a cause of limb malformations in their babies. In this study, reported by Mellin and Katzenstein (1962), 46 cases of malformed babies born in Germany in 1959 and 1960 were identified and their mothers' thalidomide exposure was compared with that of 300 mothers of normal babies, finding that 41 of the cases had been exposed to the drug compared to none of the controls.

To summarise, cross-sectional, cohort and case-control studies are distinguished only by the method of selection of subjects. In a cross-sectional study subjects are selected from a target population without regard to exposure or outcome. In a cohort study, the subjects are selected before they experience the outcome, and different selection criteria may be used for different exposure categories. In contrast, a case-control study includes some subjects who have experienced the outcome (cases) together with others who have not (controls), and different selection criteria may be used for the cases and the controls.

Experimental Studies

As noted earlier, a study is experimental (as distinct from purely observational) if the investigator has some control over a determinant of interest. Experimental studies are necessarily cohort studies: a case-control study cannot be experimental because the subjects are selected after the subjects have already been exposed to the determinant, so there is no possibility of control.

Experimental studies are concerned with investigating treatments, such as therapies for cancer patients, or interventions, such as screening and health promotion studies. These studies are classified by various factors including the type of subjects and the size of the study. The study is called a *clinical trial* if the subjects are hospital or doctors' patients, whereas the term *field trial* is used if the study involves subjects in the community at large. Thus clinical trials usually involve patients who have some disease or condition, and the objective is to investigate and compare treatments for this condition. On the other hand in field trials the subjects usually are free of disease at the time of selection, and the objective is to compare strategies or treatments for prevention.

Clinical trials are themselves classified according to phases of development of new treatments. A *phase I* trial is concerned with evaluating the safety of a proposed new treatment, whereas a *phase II* trial attempts to discover whether a treatment has any benefit for a specific outcome.

A *phase III* trial is used to compare a new treatment that has been shown to be effective with a control treatment, which could be no treatment at all. Since patients often react positively to the *idea* of a treatment (even if it is otherwise ineffective), a *placebo* treatment, that is, a treatment which looks like a real treatment but contains no active ingredient, is often used instead of no treatment. *Phase IV* trials are similar to field trials in that they involve monitoring of treatments in the community; conceptionally they differ only from field trials in the sense that they are aimed at treating subjects with some outcome whereas field trials usually focus on outcome prevention.

Experimental studies often involve randomised allocation of subjects to treatment and control groups, an idea proposed by R.A. Fisher in 1923 for comparing treatments at the Rothamsted agricultural experimental research station in Britain. The aim of randomisation is to form treatment and control groups that are as similar as possible to begin with, so that any substantial difference in outcomes observed in these groups cannot be ascribed to any factors other than the treatment effects. Randomisation ensures that the comparison groups are balanced, not just with respect to known determinants for the outcome, but also with respect to all possible risk factors.

A phase II trial that demonstrates benefit for some treatment or intervention is not necessarily conclusive, because there could be some other factor involved. Gastric 'freezing', proposed in the 1960s as a cure for duodenal ulcers, is a celebrated example of treatment that was promoted after a phase II trial, but then found to be totally ineffective after a phase III trial was conducted to compare the treatment with a placebo control. Miao (1977) has given details.

Demonstrating the absence of an association between the type of treatment and the outcome can be just as important as establishing an association. It has been established (see, for example, the National Institutes of Health, 1991) that total mastectomy is no better than less radical surgery for prolonging the survival of certain women with breast cancer, and the surgical practice of routinely recommending mastectomy in the treatment of breast cancer has now been largely discontinued.

An early large field trial was the 1954 study of the Salk vaccine for preventing poliomyelitis, described in historical terms by Meier (1972). In this trial, 400,000 schoolchildren were randomly allocated to receive the vaccine or a placebo, with the result that only 57 cases of polio occurred in the vaccinated group compared with 142 in the control group. Although the study was much more complicated than can be described by this simplistic result (as Meier explained in his article) it clearly demonstrates the vaccine's effectiveness.

4: The Credibility of a Study

Two error factors reduce the credibility of a study. These are (a) systematic error (*bias*), and (b) chance error (sampling *variability*).

Bias is a systematic distortion of the association between a determinant and an outcome due to a deficiency in the study design. It may arise (1) from poor measurement (*information* bias), (2) from the sample being unrepresentative of the target population (*selection* bias), or (3) from the differential effects of other determinants on the association of interest (*confounding*).

There are many sources of measurement bias, including faulty measuring instruments, different standards in different biochemical laboratories, errors by clinicians in diagnosing disease, bias by investigators consciously or unconsciously reporting more favourable results for treatments they believe in, biased reporting of symptoms by patients wishing to please the doctors treating them, memory lapses by subjects in case-control studies asked to recall past exposure to a risk factor (*recall* bias), lack of compliance of patients in clinical trials, and poor quality data management.

Some measurement biases may be reduced or eliminated by good study design. For example, *blinding* of investigators and subjects, so that they do not know which treatment a subject has received until after the response has been evaluated, can reduce biased reporting in clinical trials. If neither patients nor outcome evaluators know the treatment allocation, the trial is said to be *double blind*. If the evaluators know the treatment allocation but the subjects do not, the study is said to be *single blind*. While poor measurement can bias an estimate of level (such as a mean or an incidence or prevalence rate) in either direction, in some commonly occurring situations it gives rise to reduced estimates of associations, as Bross (1954) showed. Further details on this issue are given by Kleinbaum et al (1982, Chapter 12).

Selection bias has two levels. The first level occurs when the sample does not represent the target population of interest, but nonetheless constitutes a representative sample from some restricted population which is a subset of the target population. This means that although the results of the study may not be generalisable to the whole target population, the associations are still valid and may be applied to some restricted population. Such studies are said to have internal validity, but suffer from a lack of external validity. Because their subjects usually constitute a select group with tight eligibility criteria, randomised clinical trials fall into this category. For the same reason cross-sectional surveys in which the response rate is high also tend to be internally valid.

The second, more serious, level of selection bias is *differential* selection bias which arises when the selection criteria are different for groups of subjects in the study sample. Whenever a cross-sectional survey has a low response rate there is a danger of differential selection bias, because the non-responders may be atypical, or differ from the responders in some systematic way.

A cohort study also offers an opportunity for differential selection bias. As Feinstein (1985, page 286) pointed out, even the classic study of smoking and lung cancer deaths among doctors reported by Doll and Hill (1964) could have been biased. In this study questionnaires were sent to doctors inquiring about their levels of smoking, and their subsequent mortality during the next ten years was obtained from death certificates. If for any reason healthy smokers were less likely than unhealthy smokers to return the questionnaires, an inflated estimate of the risk of mortality among smokers would have resulted.

Sackett (1979) classified and labelled many biases that arise in case-control studies: such studies are particularly prone to differential selection bias, because it is difficult to obtain a control group that is a representative sample of all the non-cases in the target population. A good illustration of such selection bias is provided by three case-control studies investigating risk factors for breast cancer which were published in *Lancet* (1974). In each study, reserpine, a drug used to treat heart disease, was found to be a risk factor, and as a result of this evidence it was recommended in the journal editorial that clinicians should cease prescribing the drug to women at risk of developing breast cancer. However, subsequent more extensive studies found no evidence linking reserpine to breast cancer, and it became apparent that the apparent risk inflation associated with reserpine was due to the selection bias arising from the fact that excluding subjects with heart disease from the control groups made them atypically low in their exposure to reserpine.

Selection bias may be reduced by taking care to enumerate or classify all members of the target population and then taking a representative sample. Suppose, for example, you wish to investigate early induced abortion ('menstrual regulation') as a possible risk factor for subsequent ectopic pregnancy among women in the major city of a developing country. A case-control study could be conducted by choosing as cases all ectopic pregnant women entering hospitals in the city during a given period, with controls selected from the population of all non-ectopic pregnant women entering the same hospitals during the same period. You would need to ensure that the selected controls constituted a representative sample from the control population, and you could do this by matching each case with one or more members of the control population randomly selected from those entering the hospital on the same day.

Confounding is one of the most intriguing biases that can occur in epidemiological research. It arises whenever an outcome has two determinants which are themselves associated, and one is omitted from consideration. To get some understanding of confounding, consider the following hypothetical illustration.

Twin brothers are in their final year of high school. One wants to become a construction worker, the other a lawyer. Their father advises them to study hard, claiming that the higher their ranking at graduation, the more money they will be earning in ten years time, but the sons are sceptical. To settle the issue, it is agreed that a study will be undertaken to see if there is any correlation between graduation rank and future earnings. In this study 20 lawyers and 20 construction workers are selected, all males aged between 30 and 40, and each provides his high school graduation class percentile rank and current annual income. The sons present the data graphed as a scatter plot with a fitted straight line superimposed on the graph, as shown in the left panel of Figure 1.1.

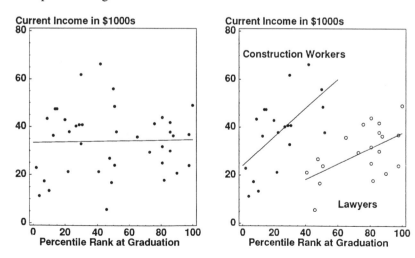

FIGURE 1.1: Illustration of Confounding

The correlation between graduation class percentile rank and current income based on this sample of 40 persons is close to 0, and the sons say 'We told you so!'

The father, not so easily convinced, computes the correlation separately in the two samples of construction workers and lawyers, obtaining a correlation coefficient of 0.62 for the construction workers and 0.56 for the lawyers. He says, triumphantly, 'these figures prove my point: it doesn't matter whether you become a construction worker or a lawyer; in each case there is a strong correlation between graduation class rank and future income'. He marks the two groups of workers in the graph with different symbols and superimposes the two lines of best fit, as in the right panel of Figure 1.1 These lines show that an increase in percentile rank of 10 points corresponds to an increase of about $6000 in annual income for the construction workers and an increase of about $3000 for the lawyers.

Confounding is one of the most common biases in epidemiological research. It can remove a valid association, as in the last illustration, or it can give rise to a spurious association. It can even reverse an association.

In our hypothetical illustration the confounding variable is the occupational group, which is strongly associated with the outcome. In fact you can see from the graph on the right that construction workers earn more than lawyers with the same percentile rank. The two determinants are also strongly associated in this case, for again it is clear from the right-hand graph that if you compare a construction worker with a lawyer having the same income, you expect the lawyer to have a higher percentile rank than the construction worker.

Confounding also distorts measures of birth and death rates in demographic research, making it necessary to make adjustments for age. Based on demographic statistics published by the United Nations (1992, pages 414–415), the crude death rate for males in Japan in 1990 was 7.3 per thousand, while that for Malaysia was only 5.3 per thousand, so it would appear that Japan had the higher mortality of the two countries. However, a truer indication of mortality is given by age-adjusted mortality rates: in each 5-year age group in 1990 the rates for Malaysia were approximately double those for Japan. The crude rate for Malaysia is lower because, in contrast to Japan, there is only a small proportion of persons in the older age groups at relatively high risk of death. Another example of age-confounding of death rates is given by Pollard et al (1990, page 73), who cite crude mortality rates of 4.65 and 8.35 deaths per 1000 for Maoris and non-Maoris in New Zealand in 1984, compared with age-adjusted rates of 10.38 and 7.87 deaths per 1000, respectively.

Although outcomes in epidemiology tend to be dichotomous

rather than continuously varying, the concept of confounding is basically as described above. Examples of confounding with dichotomous outcomes are discussed in Chapter 4.

Sampling variability arises because samples are *finite*, so even if there is no bias in a study, its conclusion is only *probably* true. The smaller the sample, the greater the extent of this sampling variability, and the less credible is the study. There are two statistical measures that are associated with the sampling variability, a *confidence interval* and a *p-value*. These quantities are defined and illustrated in the next section.

Both selection bias and confounding, as well as sampling variability, can be reduced by *matching*, a design technique which involves subdividing the treatment or exposure groups into smaller subgroups, or strata, so that all the members of a strata are homogeneous with respect to other specified determinants, such as age, which are not themselves of interest in the study. An important special case is the *matched pairs* design, where each stratum consists of just two subjects. Going further, in some situations it is feasible for each stratum to consist of just one individual: the corresponding studies are called *crossover* designs, where the subjects act as their own controls. In a crossover study, the subjects are first divided into two or more treatment groups as in a phase III trial or field study, and after a specified period of time each subject is given a different treatment to the first one. Methods for the statistical analysis of crossover trials with continuously varying outcomes have been given by Brown (1980) and by Armitage and Hills (1982), while Senn (1990) gives a more complete account which includes the case of dichotomous outcomes.

Matched studies, emphasising dichotomous outcomes, are discussed in detail in Chapter 8.

5: Sampling Variability

Assuming bias is under control, confidence intervals and p-values are measures that quantify sampling variability.

Confidence Intervals

A 95% confidence interval is an interval surrounding an estimated population characteristic (such as a prevalence or incidence rate, or a relative risk or odds ratio), and which contains the population characteristic with probability 0.95. It is thus a measure of the *precision* with which a population parameter can be determined from a study: the narrower the confidence interval, the greater the precision.

The theoretical basis for computing a confidence interval is the concept of repeating a study many times under the same conditions with different subjects. According to this theory, the subjects selected for the actual study constitute just one particular sample from a much larger (theoretically infinite) target population. (Even though the target population may not be infinite or even very large, cleverly designed sampling schemes can still give rise to much larger numbers of possible samples.)

The next step in the process of computing a confidence interval involves modelling the distribution (called the *sampling* distribution) of the parameter estimates based on the repeated studies. It is conventional to use the normal distribution (or some convenient transformation of it) as the model, and the width of the confidence interval is a multiple of the distribution's standard deviation.

As an illustration, consider a cross-sectional study involving 360 subjects in which the outcome D is dichotomous and there is just a single dichotomous exposure E, giving rise to the following contingency table:

		E	$\overline{E}$
	D	75	77
Outcome	$\overline{D}$	84	124

Exposure

Estimates of the odds of outcome in the exposed and non-exposed groups are thus $O_1 = 75/84$ and $O_2 = 77/124$, giving an estimate $(75/84)/(77/124) = 1.44$ for the odds ratio (see page 5).

Suppose now that the study is repeated, or simulated, 500 times, each time giving a different sample of 360 subjects. Table 1.1 shows the contingency tables resulting from the first two repetitions, together with that obtained from the actual study, shown on the left.

	E	$\overline{E}$			E	$\overline{E}$			E	$\overline{E}$
D	75	77		D	95	74		D	83	71
$\overline{D}$	84	124		$\overline{D}$	74	117		$\overline{D}$	82	124
	Study				Repetition 1				Repetition 2	
	(OR = 1.44)				(OR = 2.03)				(OR = 1.77)	

TABLE 1.1: Results from three Repetitions of a Study with 360 Subjects

The first repetition of the study gives an estimated odds ratio 2.03, and the second repetition gives an estimate 1.77. The distribution of the odds ratios from the 500 samples is shown as a histogram in the upper panel of Table 1.2. This histogram gives an indication of how the estimates of the odds ratio vary from one sample to another. They range from 0.6 to 3. The odds ratio in the target population is actually 1.5.

As you can see from the histogram, the distribution of odds ratios is slightly skewed to the right, but this skewness is removed by taking logarithms. A histogram of the natural logarithms of the 500 odds ratio estimates is shown in the lower panel of Figure 1.2. This frequency distribution is centred near 0.4, corresponding to an odds ratio of exp(0.4), which is close to 1.5.

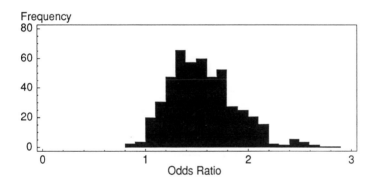

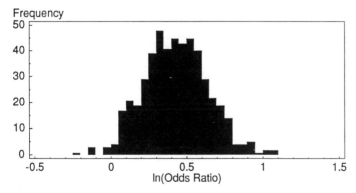

FIGURE 1.2: Histograms of Odds Ratios from 500 Repetitions of a Study

The distribution of the logarithm of an odds ratio, such as that shown in the lower panel of Figure 1.2, may be approximated by a normal distribution with mean equal to the mean of the target population, $\ln(OR)$, say, and with some standard deviation, SE, say. (By convention, the standard deviation of the distribution of estimates generated by repeated sampling from a target population is called the *standard error*, and is denoted by *SE*.)

Now it is a property of the normal distribution that 95% of its data are within 1.96 standard deviations of the mean, as shown in Figure 1.3, where the shaded tail areas represent 5% of the distribution. Taking the sampling distribution of the log odds ratios to be normal, it follows that 95% of the studies should give an estimated log odds ratio between $\ln(OR) - 1.96\,SE$ and $\ln(OR) + 1.96\,SE$. Equivalently, if you create an interval stretching from $1.96\,SE$ on either side of any given estimate T, it is 95% probable that this interval $(T - 1.96\,SE, T + 1.96\,SE)$ contains the target population mean $\ln(OR)$. Figure 1.3 shows estimates T_1 and T_2 and 95% confidence intervals for two studies: in the first, the estimate T_1 is within the central 95% probability region and its confidence interval contains the population mean: however, the second estimate T_2 is just below the 95% probability region so its confidence interval does not contain the population mean.

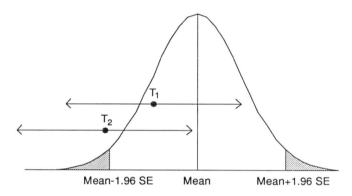

Mean-1.96 SE   Mean   Mean+1.96 SE

FIGURE 1.3: Normal Distribution and two 95% Confidence Intervals

Figure 1.4 shows the 95% confidence intervals for the odds ratio based on the first 40 repetitions of the study we have been considering. Since the odds ratios are more symmetrically distributed if they are re-expressed as logarithms, the intervals graphed in Figure 1.4 are found by first applying the formula $(T - 1.96\,SE, T + 1.96\,SE)$ to the natural logarithms of the odds ratios, and then exponentiating the results. This explains why the 95% confidence intervals graphed in Figure 1.4 are not symmetric.

Note that in three of the studies (Studies 4, 9 and 18) the confidence interval does not contain the population odds ratio of 1.5. Now on average you would expect two (5%) of the 40 95% confidence intervals to exclude the population parameter. Allowing for sampling fluctuations, this result is reasonable.

Study

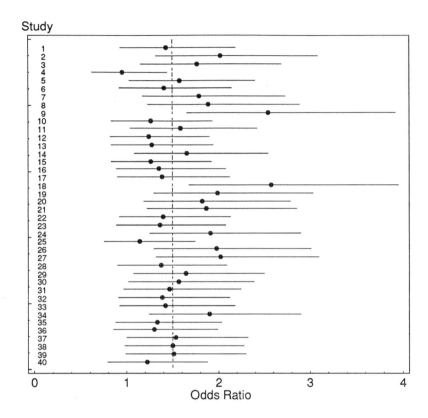

FIGURE 1.4: 95% Confidence Intervals for an Odds Ratio (value 1.5)

To get a confidence interval for a population parameter based on the results from a single study, you need to know the standard error *SE*, or at least have a reliable estimate for this quantity. Some formulas for standard errors arising in particular situations are given in the following chapters. If there is no bias, statistical theory states that the standard error decreases with the square root of the sample size. In particular this means that you need to quadruple the sample size in order to reduce the width of the confidence interval by a factor of 2, and thus double the precision of the estimate. Symbolically, this result may be expressed as

$$SE = \frac{c}{\sqrt{n}} \qquad (1.1)$$

where c is a constant and n is the sample size.

A confidence interval need not be a 95% confidence interval. Wider intervals, such as 99% confidence intervals, are preferable when many parameters are being estimated from the same study. Giving higher probability confidence intervals reduces the risk that any interval will not contain the corresponding population parameter.

P-values

A confidence interval is a measure of sampling variation associated with an estimate of a population parameter. A *p*-value is more complex, because it also involves a null hypothesis. A *null hypothesis* is a statement that a target population parameter has a particular value. Usually this value is indicative of no difference or association, for example, that a risk difference is 0 or an odds ratio is 1. A *p-value* is an indication of the plausibility of a parameter estimate with respect to a null hypothesis. It is the probability of obtaining (in repeated similar studies) an estimate as extreme as, or even more extreme than, the one obtained in the study, *assuming the null hypothesis is true*. It is usually represented as a *tail area* of a standard distribution such as a normal or chi-squared distribution.

To illustrate these concepts, consider the first 10 of the simulated studies graphed in Figure 1.4, and suppose that the null hypothesis is that the odds ratio is 1, corresponding to no association. The 95% confidence intervals are graphed in Figure 1.5, with the odds ratio corresponding to the null hypothesis represented by the dotted line. The *p*-values associated with this null hypothesis are also listed.

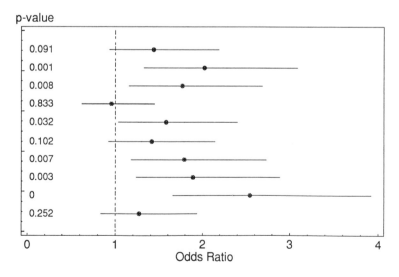

FIGURE 1.5: P-values for the First Ten Studies Graphed in Figure 1.4

The estimated odds ratio obtained from the first study is 1.44, suggesting that the odds ratio in the target population is different from 1 (the value specified by the null hypothesis). The p-value is the likelihood that a study based on data sampled from a target population with odds ratio 1 will give an *estimated* odds ratio at least as extreme as 1.44. This likelihood turns out to be 0.091. The probability is *two-tailed*, since it also includes the possibility of getting a discrepancy in the opposite direction. In other words, 0.091 is the likelihood that a study will give an odds ratio greater than 1.44 or less than 0.69 (the reciprocal of 1.44).

Similarly, the odds ratio obtained from the second study (2.03) corresponds to a p-value of 0.001, which means that if the population odds ratio is 1, the likelihood that a study will give an estimated odds ratio greater than 2.03 or less than its reciprocal is 0.001. Study 3 also gives a small p-value, indicating that the odds ratio estimated from the study is unlikely to have occurred if the null hypothesis is true.

Small p-values provide evidence against the null hypothesis being tested. Given that the 10 studies graphed in Figure 1.5 were actually generated from a population having an odds ratio of 1.5 rather than 1, it is not surprising that most (9 out of 10) of these studies give odds ratios significantly greater than 1, providing evidence against the null hypothesis. Note, however, that Study 4 gives an odds ratio less than 1, corresponding to a relatively large p-value (0.833), and Study 10 also gives a p-value that is not particularly small.

If a p-value is very small, it is reasonable to *reject* the null hypothesis. The justification for such a decision is that the data obtained from the study are inconsistent with the null hypothesis, and thus it is more reasonable to conclude that the null hypothesis is false than that a very extreme sample has arisen. By scientific convention, a p-value smaller than 0.05 is regarded as *statistically significant*, meaning that it provides sufficient evidence to reject the null hypothesis.

Although a small p-value is evidence against a null hypothesis, a relatively large p-value needs to be supported by a reasonably large sample before it provides evidence in favour of a null hypothesis. This point is discussed further in the next section.

As Figure 1.5 shows, p-values and confidence intervals are related: four of the ten studies in this graph have 95% confidence intervals that intersect the dotted line corresponding to the value of the parameter specified under the null hypothesis, and these four studies are the ones with p-values greater than 0.05. This correspondence may be explained in terms of *z-scores*, as follows.

A z-score is obtained by taking the difference between a parameter estimate and its value specified under the null hypothesis, and dividing this difference by the standard error. Thus if T is the estimate of a parameter obtained from a study and μ_0 is the null hypothesis value and SE_0 its standard error, the z-score is given by the formula

$$z = \frac{T - \mu_0}{SE_0} \qquad (1.2)$$

The p-value is now computed from the z-score as the tail area of a distribution, and if the sample size is reasonably large, this distribution is approximately standardised normal, as shown below in Figure 1.6. The p-value, by definition the likelihood of getting an estimate more extreme than the observation, is thus the total area under the standardised normal distribution in the two tails (1) to the left of $-|z|$ and (2) to the right of $|z|$, where $|z|$ is the absolute value of z.

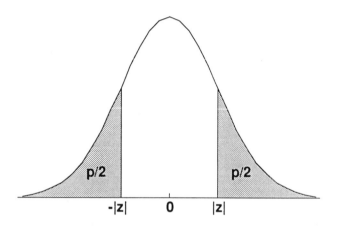

FIGURE 1.6: Two-tailed p-value from Standardised Normal Distribution

It is a property of the standardised normal distribution that if z is 1.96, the total area in the two tails is exactly 0.05, giving a p-value 0.05. Referring to Equation (1.1), when $z = 1.96$, $(T-\mu_0)/SE_0 = 1.96$, that is, $T-1.96\,SE_0 = \mu_0$. Since the 95% confidence interval extends by an amount equal to 1.96 times the standard error on either side of the estimate T, in this case the lower end of the 95% confidence interval for the parameter is equal to its value specified under the null hypothesis. Similarly, if z is greater than 1.96, the p-value is smaller than 0.05, and the lower end of the 95% confidence interval exceeds the null value of the parameter. The p-value is also smaller than 0.05 when z is less than -1.96, and in this case the upper end of the 95% confidence interval is below the null value. However, if z is between -1.96 and 1.96, the p-value is greater than 0.05, and the 95% confidence interval includes the null value (see Figure 1.3).

The duality between p-values and confidence intervals as described above is only valid, strictly speaking, when the standard error used in the confidence interval, denoted by SE, is the same as the standard error used for computing the p-value, SE_0. Otherwise the correspondence is only approximate.

6: Statistical versus Clinical Significance

Type I and Type II Errors

These errors relate to a null hypothesis. The *type I error* is the probability of rejecting a null hypothesis when it is true. The *type II error* is the probability of failing to detect a *worthwhile* effect, that is, an effect considered to be of medical importance.

The type I error is conventionally taken to be 0.05 (the criterion for statistical significance). This corresponds to a *specificity* (probability of a true negative result) of 0.95. However, the type I error should be taken to be smaller than this conventional level (for each hypothesis) if several hypotheses are being tested in the same study, thus ensuring that the overall type I error rate is controlled. Thus if there were five null equally ranked hypotheses being tested in a study, you could specify the type I error for each as 0.01. But if there were five null hypotheses with one of overriding importance, it would be more reasonable to allocate a type I error of 0.03 to the important one and 0.005 to each of the other hypotheses.

The type II error is the probability that a study will give a non-significant result when a worthwhile effect exists, that is, when the null hypothesis is false. It is thus the probability of failing to detect a worthwhile effect. The *power* or *sensitivity* of a study is the complement of the type II error, so the power is the probability of detecting an effect which is of clinical importance. The power depends on the magnitude of the effect considered worthwhile, the sample size, the variation in measurements, the study design, and the type I error.

For example, a cohort study involving 41 subjects reported by Rossing et al (1993) found no association between albuminuria and glomerular filtration rate decline, after adjusting for differences in diastolic blood pressure. However, if a worthwhile association is taken as a correlation coefficient of 0.4 (or more), the power of this study turns out be only 40%, so it is quite possible that the study failed to detect a worthwhile effect: the study was *inconclusive*.

The concept of statistical power is illustrated in Figure 1.4, where 95% confidence intervals for 40 similar simulated studies based on a target population odds ratio of 1.5 are depicted. It turns out that

20 of the simulated 40 studies – those where the confidence interval does not contain 1 – give a significant p-value (that is, $p < 0.05$) for testing the null hypothesis that the population odds ratio is 1. The estimated power is thus 0.5.

Taking the type I error to be 0.05 means that 5% of true null hypotheses are going to be falsely rejected. As a consequence, there is a danger in undertaking studies in which there is a small likelihood that the effect is worthwhile, even if the power is high. To see this, suppose that the likelihood that the effect is worthwhile is only 10% and the power of a study is 50%. Out of 200 such studies, 180 are investigating no worthwhile effect in the target population but you expect nine (5%) of them to give significant results, whereas you expect 10 (50%) of the 20 studies investigating a worthwhile effect to detect it. So nearly half of the significant findings reported will be false positives. Note that increasing the power does not solve the problem: even if the power of each study is 100% almost one third of the significant findings reported will be incorrect.

It is important to understand the difference between 'statistical' and 'clinical' significance. According to scientific convention, a finding is statistically significant if the p-value associated with the null hypothesis of interest is smaller than 0.05. On the other hand, a result is clinically important if the confidence interval for the parameter of interest differs from the null value by a worthwhile amount.

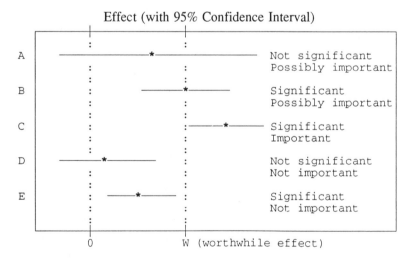

FIGURE 1.7: Statistical Significance versus Clinical Importance

Figure 1.7 illustrates the different kinds of results that can eventuate from a study which measures a parameter in a target population. We assume that the value of the parameter is 0 under the null hypothesis, and a value of W is considered to be a worthwhile difference from the null value.

In Situation A the confidence interval covers the null value, corresponding to a non-significant p-value. However the worthwhile effect is also contained in the confidence interval, so the result could be important. In this case the study is inconclusive.

In Situation B the 95% confidence interval excludes the null value, so the result is statistically significant. However, the study does not necessarily indicate a worthwhile effect, since the confidence interval includes W, and the effect could thus be not worthwhile.

In Situation C there is both statistical significance and clinical significance, because the confidence interval is completely above W. A conclusive result is also obtained in Situation D, where the study has found a non-significant p-value and no worthwhile effect.

The final possibility is shown as Situation E, where a statistically significant finding is not clinically important.

As a consequence, if a study is based on a small sample, a relatively large p-value (null result) is not informative. But if a study is large, a small p-value is meaningless if the effect is not important. Another way of looking at this is to consider what happens in Situation A when the sample size is gradually increased, assuming that the estimate of the effect remains the same. The length of the confidence interval will gradually decrease and at a certain point it will not intersect the null value (as in Situation B). As the sample size increases further, the confidence interval will not intersect the worthwhile effect (as in Situation E).

Thus the study should be large enough to detect a worthwhile effect, but not so large that it can detect an unimportant effect.

Summary

Epidemiological research involves investigating associations between disease outcomes and their putative determinants in human populations. It involves measuring the relevant variables on individuals selected as subjects for a study and using statistical analysis to draw conclusions which apply to an appropriate target population.

Measurement

Outcome measurements are often dichotomous, such as disease presence or absence (measured by prevalence or incidence), but could also be multivalued or continuous. Determinants include genetic, demographic, environmental, occupational and behavioural risk factors, prevention and disease treatment. Risk differences, relative risks and odds ratios measure associations between determinants and outcomes.

Study Types

Studies are observational (cross-sectional, cohort, and case-control studies) or experimental (clinical trials and field trials). Observational studies are classified according to the method of selection of subjects.

The Credibility of a Study

Bias (information bias, selection bias, and confounding) and sampling variability (due to finite study size) reduce the credibility of a study.

Sampling Variability

Sampling variability is reflected in confidence intervals (containing the population parameters of interest with specified probability) and *p*-values. A *p*-value measures the plausibility of a null hypothesis, being the probability of observing the data when the null hypothesis is true.

Statistical versus Clinical Significance

The type I error is the probability of rejecting a true null hypothesis (conventionally 5%) while the type II error is the probability that a study will fail to detect a true worthwhile effect. Too small a study is likely to fail to detect a worthwhile (clinically important) effect, but too large a study could easily detect an effect that is not worthwhile.

Exercises

The following abstracts are taken from recent epidemiological journals. In each case try to answer the following questions using the information provided in the abstract.

(a) Describe the target population and the sample used in the study.
(b) What is the question of interest?
(c) What are the determinant and outcome variables?
(d) What is the study type?
(e) What possibilities for bias exist?
(f) Describe the covariates. Is there confounding?
(g) Where does the study fit in Figure 1.7?
(h) Could a different study design have been used? Explain.
(i) Are the results adequately reported?

Nitrous Oxide and Spontaneous Abortion in Female Dental Assistants

Andrew S. Rowland, Donna D. Baird, David L. Shore, Clarice R. Weinberg, David A. Savitz, and Allen J. Wilcox

The relation between anesthetic gas exposure and spontaneous abortion remains unresolved. We examined the effect of nitrous oxide on spontaneous abortion among female dental assistants. Questionnaires were sent to 7000 dental assistants aged 18–39 years who were registered in California in 1987; 4856 (69%) responded. Analysis was based on 1465 respondents whose most recent pregnancy was conceived while working full time. Women were asked how many hours a week they worked with nitrous oxide during this pregnancy and whether the excess gas was scavenged (vented). Relative risk of spontaneous abortion (through week 20) was calculated using a person-week model. This allowed women with current pregnancies (13%) or induced abortions (10%) to be included for appropriate time periods of risk. A total of 101 pregnancies (7%) ended as spontaneous abortions. An elevation in risk of spontaneous abortion was seen among women who worked with nitrous oxide for 3 or more hours per week in offices not using scavenging equipment (relative risk = 2.6, 95% confidence interval 1.3–5.0, adjusted for age, smoking, and number of amalgams prepared per week), but not among those using nitrous oxide in offices with scavenging equipment. This relation changed little when analyses were restricted to confirmed pregnancies or examined for several types of potential bias. Scavenging equipment appears to be important in protecting the reproductive health of women working with nitrous oxide. *Am J Epidemiol* 1995; 141:531-8.

The sleep patterns of normal children

Kenneth L. Armstrong, Robyn A. Quinn and Mark R. Dadds

Objectives: To determine the range of sleep behaviour of normal children to age 38 months and to ascertain the level of parents' problems associated with their child's sleep behaviour.

Design and setting: A cross-sectional survey by questionnaire of parents presenting with their children for routine well-child checks at child health centres, mobile clinics, flying doctor clinics and home visits throughout Queensland. Of 3383 questionnaires distributed 3269 (96.5%) were returned.

Main outcome measures: 1. Sleep frequency and duration, settling procedures, time taken to settle at night, age when child first slept through the night and number of night-time wakenings requiring parental intervention. 2. Parents' problems with their child's sleep behaviour.

Results: There is a wide range of normal childhood sleep behaviour. Circadian rhythm is not well established until four months of age. Daytime sleep becomes less regular with increasing age. Frequent night-time wakening is common from 4 to 12 months. Night-time settling requires more parental input from 18 months. A large proportion of parents (28.6%) have a problem with their child's sleep behaviour.

Conclusions: Parents require information from health care providers about the wide range of normal childhood sleep patterns. This information can help prevent misdiagnosis, inappropriate medication use, child abuse and parental depression when children's sleep patterns are a problem.

Med J Aust 1994; 161: 202–206.

Mortality from cardiovascular disease among interregional migrants in England and Wales

D.P. Strachan, D.A. Leon, B. Dodgeon

Objective: To investigate the extent to which geographical variations in mortality from ischaemic heart disease and stroke in Britain are influenced by factors in early life or in adulthood.

Design: Longitudinal study of migrants.

Subjects: 1% sample of residents in England and Wales born before October 1939 and enumerated at the 1971 census (the Office of Population Censuses and Surveys' longitudinal study).

Main outcome measure: 18 221 deaths from ischaemic heart disease and 9899 deaths from stroke during 1971-88 were analysed by areas of residence in 1939 and 1971. These included 2928 deaths from ischaemic heart disease and 1608 deaths from stroke among individuals moving between 14 areas defined by the major conurbations and nine standard administrative regions of England and Wales.

Results: The southeast to northwest gradient in mortality from ischaemic heart disease was related significantly to both the 1939 area ($\chi^2 = 6.09$, df = 1) and area in 1971 ($\chi^2 = 5.05$, df = 1). Geographical variations in mortality from stroke were related significantly to the 1939 area ($\chi^2 = 4.09$, df = 1) but the effect of area in 1971 was greater ($\chi^2 = 8.07$, df = 1). The effect of 1971 area on mortality from stroke was largely due to a lower risk of death from stroke among individuals moving into Greater London compared with migrants to the rest of the south-east region ($\chi^2 = 4.54$, df = 1).

Conclusions: Geographical variations in mortality from cardiovascular disease in Britain may be partly determined by genetic factors, environmental exposures, or lifestyle acquired early in life, but the risk of fatal ischaemic heart disease and stroke changes on migration between areas with differing mortality. The low risk of death from stroke associated with residence in Greater London is acquired by individuals who move there.

References

Altman, D.G. (1991): *Practical Statistics for Medical Research.* Chapman & Hall. London.

Armitage, P. & G. Berry (1994): *Statistical Methods in Medical Research* (3rd ed.). Blackwell Scientific Publications. Oxford.

Armitage, P. & M. Hills (1982): The two-period cross-over trial, *The Statistician*, **31**, pages 119–131.

Breslow, N.E. & N.E. Day (1980): *Statistical Methods in Cancer Research: Volume I – The Analysis of Case-Control Studies*. IARC. Lyon.

Breslow, N.E. & N.E. Day (1987): *Statistical Methods in Cancer Research: Volume II – The Design and Analysis of Cohort Studies*. IARC. Lyon.

Bross, I.J.D. (1954): Misclassification in 2 x 2 tables, *Biometrics* **10**, pages 478–486.

Brown, B.W. Jr. (1980): The crossover experiment for clinical trials, *Biometrics* **36**, pages 69–70.

Dawber, T.R. (1980): *The Framingham Study: The Epidemiology of Atherosclerotic Disease*. Harvard University Press. Cambridge.

Doll, R. & A.B. Hill (1964): Mortality in relation to smoking: ten years observations of British doctors, *British Medical Journal*, **1**, pages 1399–1410 and 1460–1467.

Feinstein, A.R. (1985): *Clinical Epidemiology*. W.B. Saunders. Philadelphia.

Hill, A.B. (1951): The clinical trial, *British Medical Bulletin* **7**, pages 278–287.

Kleinbaum, D.G., L.L. Kupper and H. Morgenstern (1982); *Epidemiologic Research: Principles and Quantitative Methods*. Van Nostrand Reinhold. New York.

Lancet (1974): Editorial, **2**, page 701.

Louis, P.C.A. (1834): An essay on Clinical Instruction. Translated by Martin P. Highley, London, pages 26–27.

Louis, P.C.A. (1836): Researches on the Effects of Bloodletting in some Inflammary Diseases, and on the Influence of Tartarized

Antimony and Vesication in Pneumonius. Translated by C.G. Putnum, Hilliard Gray, Boston.

Meier, P. (1972): The biggest public health experiment ever: the 1954 field trial of the Salk poliomyelitis Vaccine, in *Statistics: A Guide to the Unknown* (edited by J.M. Tanur), pages 2–13. Holden Day. San Francisco.

Mellin, G.W. & M. Katzenstein (1962): The saga of thalidomide: neurotrophy to embryopathy, with case reports of congenital anomalies, *New England Journal of Medicine*, **267** (23), pages 1184–1193; **267** (24), pages 1238–1244.

Miao, L.L. (1977): Gastric freezing: An example of the evaluation and medicine therapy by randomized clinical trials, in J.P. Bunker, B.A. Barnes & F. Mosteller (eds) *Costs, Risks & Benefits of Surgery*. Oxford. New York.

National Institutes of Health (NIH) Consensus Development Conference Statement: Treatment of Early Stage Breast Cancer (1991): *Journal of the American Medical Association* **265**, pp 391–395.

Pollard, A.H., Farhat Yusuf & G.N. Pollard (1990): *Demographic Techniques* (3rd ed.). Pergamon. Sydney.

Rossing, P., E. Hommel, U.M. Smidt & H-H Parving (1993): Impact of arterial blood pressure and albuminuria on the progression of diabetic nephropaphy in IDDM patients, *Diabetes*, **42**, pages 715–719.

Sackett, D.L. (1979): Bias in analytic research, *Journal of Chronic Diseases* **32**, pages 51–63.

Schlesselman, J. (1982): *Case-Control Studies*. Oxford University Press. New York.

Senn, S. (1990): *Cross-over Trials in Clinical Research*. John Wiley & Sons. New York.

Snow, J. (1855): *On the mode of communication of cholera*. London, Churchill (Reprinted in *Snow on cholera: a reprint of two papers*. New York. Hafner, 1965).

United Nations (1992): *Demographic Yearbook 1991*. New York.

2

STATISTICAL METHODS I

1: Introduction

Statistical methods are used to analyse data and thus gain information from designed studies, and to help answer questions of interest by providing confidence intervals and p-values. Statistical models are used to provide a more rigid framework for data analysis and to assist with interpretations of results. In this chapter and the next we give an outline of the simplest statistical methods for analysing data from epidemiological studies. These methods apply to situations in which there is a single outcome factor and a single determinant and no other covariates.

Different statistical methods have been developed for analysing different data types. For example if both the outcome and the determinant are dichotomous, the data may be represented as a two-by-two contingency table of frequencies, and it is of interest to compare the two resulting proportions. In this case an odds ratio may be used as a measure of association. If the variables are classified into more than two categories the data may still be represented as a two-way contingency table of counts, and summarised using multiple proportions and odds ratios. For continuous outcome data it is conventional to use means and differences of means rather than proportions and odds ratios to measure associations of interest, and correlation and regression methods are most convenient for handling data where both the outcome and the determinant are continuous.

The chapters in this book are classified largely by the complexity of the determinant variable. In this introductory chapter we consider situations in which there is a single dichotomous determinant. The statistical methods for analysing such data include chi-squared tests for comparing proportions and t-tests for comparing means. In Chapter 3 these methods are extended to allow for determinants classified into several categories, and they include one-way and two-way analysis of variance and simple regression. The remaining chapters allow for continuous and multivariate determinant variables, and include Mantel–Haenszel methods for handling confounders, logistic and Poisson regression, survival analysis, and polytomous regression.

2: Two-by-two Tables

The data in the following table are taken from a randomised trial reported by Hampton and Hill (1978). The study sample comprised 264 subjects living in a well–defined area surrounding a central hospital, who were visited by a doctor and suspected of having suffered an acute myocardial infarct but who did not need immediate hospitalisation. They were randomly allocated to either (a) home management or (b) hospital care. The outcome variable here is death within a further six weeks.

| | | Treatment | |
		Home	Hospital
Outcome	Died	17	14
	Survived	115	118
		132	132

TABLE 2.1: Results from a Randomised Trial

The authors concluded 'for the majority of patients to whom a general practitioner is called, because of suspected infarction, hospital admission confers no real advantage'. Based on the data given, is this conclusion justified? In this section this question is addressed using statistical methods.

The data arose from a cohort study, because the subjects were selected from a target population and observed for a specified time during which an outcome event of interest could occur. In this case the target population consists of patients who had suffered a mild heart attack (more precisely, patients who were suspected of having had an acute myocardial infarct but who did not need to be sent to hospital), the event of interest is death, and the period of observation is six weeks.

In this case the research question is 'should a patient be sent to hospital after suffering a mild heart attack, or should the patient stay at home and be looked after by a nurse?'

To answer this question properly several issues need to be considered. The relative costs of the two alternatives should be taken into account. Staying in a hospital is expensive compared to home care. Also the home usually (but not always) provides a happier environment than a hospital. However, a person who has suffered a heart attack is at risk of dying from a subsequent more serious heart attack, and the benefits of staying at home may be outweighed if this risk is substantially reduced by going to hospital.

The risks of death for the two alternatives may be compared by undertaking an experiment in which the patients in the study are allocated to two treatment groups, enabling the outcomes to be compared. If the proportion of patients dying after home care turns out to be greater than the proportion dying after hospital treatment, then you would prefer to send a patient to hospital. But if these proportions are much the same, then you would prefer to keep the patient at home.

An ethical question arises. Is it justifiable to conduct an experimental study in which one of the treatments may substantially increase the risk of death? Let us consider this question.

It is not ethical to do an experimental study in which you *know* that one of the treatments is riskier than another. However, if the treatments are not known in advance to have differing risks, then it is ethical to do an experimental study provided the patients give their consent. The purpose of doing the study is to find out if one treatment is better. If, as a result of a study, you find out that one treatment has a greater risk, then the result should be reported so that further studies need not be undertaken. However, if no difference is found it may be because your study has failed to detect a difference that really exists in the target population. This could happen if the number of subjects in your study is too small. You should still report this result, even though it is inconclusive, because if your results are combined with other studies the net result may be conclusive.

Note that the study involves randomisation. Why? Because it has become accepted that randomisation is the only effective method for ensuring that treatment groups are similar to begin with. If the treatment groups are different in some respects before the treatment is given, a difference observed after the treatment may not be due to the treatment. Having groups that are similar is the only way of ensuring that any difference observed is associated solely with the treatment. Balancing the treatment groups with respect to observable factors such as age, sex, and health status, is not enough, because there may some unobserved (genetic, or environmental) factor that is associated with the treatment and which for some reason is present in only one treatment group.

Confidence Intervals

The subjects in the study constitute a sample from the target population of interest. If the study were to be repeated, a different sample would be obtained and the results would be different. Statistical methods are needed to allow for sampling error.

The extent of the sampling error is determined by several

factors, including the sample size, the type of study, the method of data collection, and the distribution of the measurements taken from the subjects in the study.

Sampling errors are usefully summarised by the standard errors of the statistics (such as means and proportions) obtained from the study. In the study of home-versus-hospital care, the statistics of interest are the proportions of patients dying in the two groups. The observed proportions are 17/132 (12.9%) in the home-care group and 14/132 (10.6%) in the hospital-treated group. These are estimates of the risks of dying within six weeks after the respective treatments, in the target population. The standard error is a measure of the probable range of variation of a statistic over repeated samples of the same size. This probable range is based on statistical theory, which shows that approximately 68% of values of an observed statistic, based on independent samples of the same size, are within one standard error of the population parameter. This approximation becomes more exact as the sample size increases.

Statistical theory provides a formula for the standard error of a proportion, namely

$$SE(p) = \sqrt{\frac{p(1-p)}{n}} \qquad (2.1)$$

where p is the observed proportion and n is the sample size from which this proportion is calculated. Applying this formula to the observed proportions in the home-versus-hospital care study, you get standard errors of 0.029 for the proportion dying in the home treated group and 0.027 for the proportion dying in the hospital treated group.

Standard errors may be used to obtain confidence intervals for parameters in the target population. (Recall that a confidence interval is an interval based on the sample statistic that contains the population parameter with specified certainty.) From the definition of the standard error it follows that the interval spanning one standard error on either side of a sample statistic is a 68% confidence interval. Conventionally, 95% confidence intervals are preferred when reporting results from studies, and according to statistical theory the interval spanning two standard deviations on either side of a sample statistic approximates a 95% confidence interval for the population parameter. This recipe gives 95% confidence intervals of $(12.9 - 2 \times 2.9, 12.9 + 2 \times 2.9)$, or $(7.1, 18.7)$ for the percentage risk of death in the home-care group, and $(10.6 - 2 \times 2.7, 10.6 + 2 \times 2.7)$, or $(5.2, 16.0)$ for the percentage risk of death in the hospital care group.

This confidence interval is based on what is known as the *two standard error rule* and is not exact, even asymptotically. More precise 95% confidence intervals are obtainable using relevant statistical theory. For large samples, a better approximation to a 95% confidence interval results by taking 1.96, rather than 2, standard errors on each side of the estimate. However, for simplicity 95% confidence intervals are often calculated by taking two standard errors on each side of an appropriate estimate of the population parameter, and these are only approximately valid.

When comparing two population parameters it is useful to report a confidence interval for the difference (or perhaps the ratio) of the two parameters. Using statistical theory it may be shown that the standard error of the difference between two proportions is given (again approximately) by the formula

$$SE(p_1 - p_2) = \sqrt{\frac{p_1(1-p_1)}{n_1} + \frac{p_2(1-p_2)}{n_2}} \qquad (2.2)$$

where p_1 and p_2 are the two observed proportions and n_1 and n_2 are the sample sizes from which they are calculated. Using the data from our study, the observed difference in proportions is $12.9 - 10.6 = 2.3\%$, and the standard error of this difference, using the above formula, is 0.039. A 95% confidence interval for the difference in risks in the target population is thus $(2.3 - 2 \times 3.9, 2.3 + 2 \times 3.9) = (-5.5, 10.1)$.

Based on this 95% confidence interval, for persons who suffered heart attacks in the target population, the increase in the risk of death associated with home care rather than hospital care could have been as large as 10.1% or as small as −5.5% (a decrease).

Note that the risks may be compared in other ways than by looking at their difference: you could look at the relative risk, that is, the ratio of the risk of dying after home care to the risk of dying after hospital care. In medical studies it is actually more common to use the odds ratio to compare two risks. The odds of an outcome is defined in terms of the risk p as the ratio $p/(1-p)$, which is the ratio of the probability of occurrence to the probability of non-occurrence. If the risk is small, the odds ratio is a close approximation to the relative risk. An odds ratio (or relative risk) equal to 1 corresponds to equal risks in the two groups being compared.

The home-versus-hospital study gave an observed relative risk $0.129/0.106 = 1.22$, and an observed odds ratio $0.148/0.119 = 1.24$.

A 95% confidence interval for an odds ratio may be obtained from a formula for the standard error of the natural logarithm of an observed odds ratio. The formula is

$$SE(\ln OR) = \sqrt{\frac{1}{a} + \frac{1}{n_1 - a} + \frac{1}{b} + \frac{1}{n_2 - b}} \qquad (2.3)$$

where a and b are the numbers of outcomes and n_1 and n_2 are the respective sample sizes in the two groups. The confidence interval for the odds ratio is obtained by first using this formula to get the confidence interval for the logarithm of the odds ratio, and then exponentiating the values defining the interval.

The home-versus-hospital study has $a = 17$, $b = 14$, $n_1 = 132$ and $n_2 = 132$, corresponding to a standard error 0.384. The logarithm of the observed odds ratio is thus $\ln(1.24) = 0.22$, corresponding to a 95% confidence interval $(0.22 - 2 \times 0.384, 0.22 + 2 \times 0.384)$, that is, $(-0.55, 0.99)$. Exponentiating these values, the 95% confidence interval for the odds ratio itself is $(e^{-0.55}, e^{0.99}) = (0.58, 2.69)$. So you would conclude from this study of the risk of six-week mortality in the target population, the odds ratio for home care compared to hospital care is very probably between 0.58 and 2.69. This conclusion is consistent with the one based on the difference between the two proportions.

Graphical Displays

Confidence intervals may be graphed using line intervals. Figure 2.1 shows a graph containing the 95% confidence intervals for the risks of death within six weeks corresponding to the home-care and hospital-care groups of patients with heart disease.

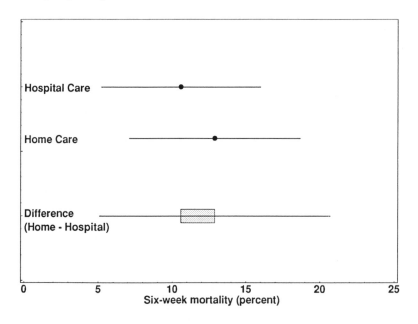

FIGURE 2.1: Results of a Study involving 264 Heart Patients

The graph also highlights the estimated difference between these proportions (shown as the shaded box) and gives the 95% confidence interval for this difference, represented as a line centred at the right-hand end of the box.

Graphing confidence intervals is a very effective way of summarising the results obtained from a study. It is clear from Figure 2.1 that the risks of death could be the same for each treatment, because the confidence intervals overlap a lot. However, the graph also shows that there could be quite a large difference between these risks, because the confidence interval for the difference (shown as the bottom line in the graph) is relatively large.

Hypothesis Testing

A confidence interval is a measure of the sampling variation in a study, giving a range of values between which the target population parameter is very likely to occur. A confidence interval thus enables an investigator to give a valid assessment of the result of a study. If the confidence interval covers a wide range the study is not very informative, and possibly inconclusive. On the other hand if the confidence interval is narrow the study is more informative.

In this section we give another way of assessing the result from a study, using a *p*-value based on testing a null hypothesis.

A null hypothesis is a statement that a population parameter of interest has a specified value. This specified value is called the null value and is related to the objective of the study. For an odds ratio the null value is conventionally taken to be 1, corresponding to equal risks of an outcome in two comparison groups. This corresponds to a null value of 0 for the difference between two population proportions.

A null hypothesis is tested by comparing an appropriate statistic based on the observed data with its null value (or more precisely the null value of the equivalent population parameter). The *p*-value is then defined as the probability that another sample would give a discrepancy as least as large as the one observed.

The greater the discrepancy between the statistic based on the study and its null value, the smaller the *p*-value. However, if the study gives a statistic close to the null value, the *p*-value is relatively large. In the extreme situation when the study statistic is equal to the null value (so that the discrepancy is 0), the *p*-value is 1, because it is certain that another study will give a discrepancy of at least 0.

A two-by-two contingency table of counts provides an illustration of hypothesis testing. Suppose that there are two possible outcomes D_1 and D_2 for independent cases classified into two exposure groups labelled E_1 and E_2, and that the observed counts in the four categories are *a*, *b*, *c* and *d*, respectively, as shown in the following table. (Note that the home-versus-hospital study provides a special case of this data layout, where in that case the observed counts are $a = 17$, $b = 14$, $c = 115$ and $d = 118$.)

An appropriate null hypothesis here is that the odds ratio is 1, which is equivalent to the statement that the probabilities of the outcomes in the two exposure groups are the same. This null hypothesis is called the independence hypothesis in a contingency table, because it assumes that the outcome is independent of the exposure.

	Exposure	
	E_1	E_2
D_1	a	b
D_2	c	d

Outcome (for rows D_1, D_2)

A p-value is obtained from Pearson's chi-squared test, which may be described as follows. The observed counts are compared with expected counts based on the null hypothesis, and obtained by multiplying together marginal row and column totals and dividing by the overall total. The discrepancies between the observed and expected counts are squared and divided by the expected counts, and these results are summed to obtain a statistic which has a chi-squared distribution with 1 degree of freedom when the null hypothesis is true. The p-value is now given by the probability of getting a chi-squared random variable greater than the observed statistic.

For the home-versus-hospital care study, the observed counts (reading across rows) are 17, 14, 115 and 118, so the expected counts are 15.5, 15.5, 116.5 and 116.5, and the standardised squared discrepancies are 0.145, 0.145, 0.019 and 0.019, which sum to 0.329. The probability that a chi-squared random variable with 1 degree of freedom exceeds 0.329 is 0.57. Thus the p-value is 0.57.

If the p-value is not small, and by convention this means larger than 0.05, there is said to be 'insufficient evidence' to reject the null hypothesis. On the other hand if the p-value obtained from a study is smaller than 0.05 the result is said to be 'statistically significant'.

For the home-versus-hospital study the correct conclusion is that the study provides insufficient evidence to reject the null hypothesis that the risks are the same in the two treatment groups.

Note that there is a duality between hypothesis tests and confidence intervals. If a 95% confidence interval does not contain the null value of a population parameter, then you expect the p-value to be smaller than 0.05 and thus the result is statistically significant. Conversely if a 95% confidence interval includes the null value of the parameter, the p-value should be greater than 0.05 and thus there will be insufficient evidence to reject the null hypothesis.

3: Multiple Outcomes

In this section the methods described in the preceding section are extended in a straightforward way to designs where the outcome is classified into more than two categories.

As an illustration of the application of these methods, consider the problem of analysing toxicities in a drug trial. In 1986 the ANZ Breast Cancer Trials Group began a study (ANZ 8614) to evaluate single agent mitozantrone (MZ) versus combination cytotoxic (CMFP) therapy in patients with advanced breast cancer. Table 2.2 gives the incidence of worst overall nonhaematologic toxicity grade by treatment for 273 subjects randomised before June 1992. The toxicity grades range from 0 (none) to 4 (very severe).

	Grade					
	0	1	2	3	4	Total
MZ	13	32	55	31	2	133
CMFP	7	30	55	34	14	140
Total	20	62	110	65	16	273

TABLE 2.2: Worst Toxicities in a Breast Cancer Trial

The null hypothesis of no association may be tested by computing a chi-squared statistic, obtained by subtracting expected cell counts from the observed counts, dividing the squares of these residuals by the expected counts, and adding the resulting components. The expected counts are obtained by dividing the products of the marginal totals by the overall total. This calculation gives results $133 \times 20 / 273 = 9.74$, $133 \times 62 / 273 = 30.20$, $133 \times 110 / 273 = 53.59$, $133 \times 65 / 273 = 31.67$ and $133 \times 16 / 273 = 7.80$ in the first row, and $140 \times 20 / 273 = 10.26$, $140 \times 62 / 273 = 31.80$, $140 \times 110 / 273 = 56.41$, $140 \times 65 / 273 = 33.33$ and $140 \times 16 / 273 = 8.20$ in the second row. These expected counts may be tabulated as follows:

	Grade				
	0	1	2	3	4
MZ	9.74	30.20	53.59	31.67	7.80
CMFP	10.26	31.80	56.41	33.33	8.20

The standardised residuals are computed as $(13 - 9.74)^2/9.74 = 1.09$, $(32 - 30.2)^2/30.2 = 0.107$, etc. These add up to 10.82, the chi-squared statistic. The p-value is the area of the upper tail of the chi-squared distribution beyond the observed statistic, where the number of degrees of freedom is one less than the number of categories into which the outcome is classified. In this case there are 5 categories, so the number of degrees of freedom is 4. The area beyond 10.82 for the chi-squared distribution with 4 degrees of freedom is 0.029, so the null hypothesis should be rejected in this case. The conclusion is that there is an association between the treatment and the toxicity grade.

The association in a contingency table may be displayed by graphing odds ratios with their 95% confidence intervals. Recall that for a 2-by-2 table having counts a and b in the first row and c and d in the second row, the odds ratio is given by $(a/c)\,/\,(b/d)$, and the standard error of its natural logarithm is $\sqrt{(1/a+1/b+1/c+1/d)}$. For a contingency table with several different outcome categories, one of these categories should be chosen as the referent category (similar to the null outcome when there are just two categories), and an odds ratio and corresponding confidence interval may then be calculated for each of the other outcome categories.

For the present example, it is reasonable to choose toxicity grade 0 as the referent category. The odds of toxicity grade 1 (compared to grade 0) is 30/7 for CMFP and 32/13 for mitazantrone, so the odds ratio is $(30/7)\,/\,(32/13) = 1.74$, with natural logarithm 0.55. The standard error of this log odds ratio is $\sqrt{(1/30+1/7+1/32+1/13)} = 0.53$, giving a 95% confidence interval $(e^{-0.51}, e^{1.62})$, or (0.60, 5.0) for the odds ratio. The odds ratio for toxicity grade 2 is $(55/7)\,/\,(55/13) = 1.86$, with natural logarithm 0.62. The standard error of this estimate is 0.51, giving the 95% confidence interval $(e^{-0.39}, e^{1.63})$, or (0.67, 5.2). The confidence intervals for the remaining log odds ratios are calculated in a similar way, and are displayed in Figure 2.2.

These confidence intervals are graphed on a logarithmic scale, making them symmetric. Base 2 rather than base 10 or natural (base e) logarithms are used, because each unit increase on the scale then corresponds to a doubling of the odds, facilitating interpretation. For example the graph shows readily that for the risk of a very severe worst toxicity (compared to none) the confidence interval for the odds ratio comparing CMFP to mitazantrone is between 2 and 64 (2^6).

You can also see from Figure 2.2 that the odds ratios comparing toxicity grades 1, 2 and 3 with grade 0 are statistically indistinguishable. The only confidence interval that does not overlap the null value is that corresponding to the 'very severe' toxicity grade. Thus the conclusion is that it is only at the most extreme grade of toxicity that there is a treatment difference.

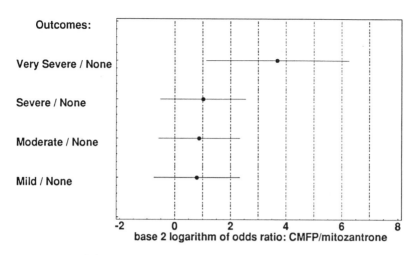

FIGURE 2.2: Worst Toxicity for 273 Cancer Patients

4: Continuous Outcomes

The first example in the preceding section involves a dichotomous outcome, namely, whether a patient who had suffered a mild heart attack was alive or dead six weeks later. In the second example the outcome has five categories corresponding to different toxicity grades. We now consider an example where the outcome is continuous.

Sternberg et al (1982) reported the results from a study involving 25 hospitalised schizophrenic patients classified as either 'psychotic' or 'nonpsychotic'. Samples of cerebrospinal fluid were taken from each patient and assayed for the dopamine b-hydroxylase (DBH) activity, with the following results (the units being nmol/(ml)(h)/(mg) of protein):

Judged nonpsychotic

0.0104   0.0105   0.0112   0.0116   0.0130   0.0145   0.0154
0.0156   0.0170   0.0180   0.0200   0.0200   0.0210   0.0230
0.0252

Judged psychotic

0.0150   0.0204   0.0208   0.0222   0.0226   0.0245   0.0270
0.0275   0.0306   0.0320

The question of interest is whether the psychotic and non-psychotic patients have different DBH levels. The two population means are compared using confidence intervals, and the null hypothesis that these means are equal is tested using a two-sample t-test.

Note that the mean DBH level of the sample of nonpsychotic patients is 0.0164, which is substantially less than the mean obtained for the psychotic patients (0.0243). Confidence intervals may be used to assess the statistical significance of the difference in these means.

The standard error of a sample mean $\bar{y}$ is given by the formula

$$SE(\bar{y}) = \frac{s}{\sqrt{n}} \qquad (2.4)$$

where s is the standard deviation of the data in the sample and n is the size of the sample. A 95% confidence interval for the population mean extends two standard errors on each side of the sample mean.

The standard deviations of the two samples of DBH levels are 0.0047 for the nonpsychotics and 0.0051 for the psychotics. Thus the corresponding standard errors from Equation (2.4) are $0.0047/\sqrt{15} = 0.0012$ and $0.0051/\sqrt{10} = 0.0016$, giving 95% confidence intervals (0.0140, 0.0188) for the nonpyschotics and (0.0211, 0.0275) for the psychotics. These confidence intervals are graphed in Figure 2.3. This graph also shows the difference between the sample means, depicted as a shaded box, and a 95% confidence interval for the difference, shown as a line interval centred at the right-hand end of the box.

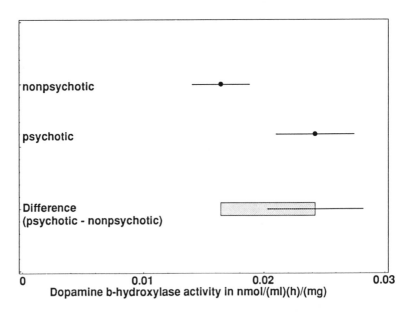

FIGURE 2.3: 25 Hospitalised Schizophrenic Patients

The confidence interval for the difference in the means is based on the formula

$$SE(\bar{y}_1 - \bar{y}_2) = s \sqrt{\frac{1}{n_1} + \frac{1}{n_2}} \qquad (2.5)$$

where s is the pooled sample standard deviation, given by

$$s = \sqrt{\frac{(n_1 - 1)s_1^2 + (n_2 - 1)s_2^2}{n_1 + n_2 - 2}} \qquad (2.6)$$

and n_1 and n_2 are the sizes and s_1 and s_2 the standard deviations of the two samples.

Equation (2.6) is based on a statistical assumption, namely that the standard deviations in the two populations are the same. If these standard deviations are too different the confidence interval for the difference between the population means is not valid. This assumption may be assessed by graphing box plots of the data in the two samples. In case you are not familiar with a box plot, here is a brief definition.

A box plot, suggested by Tukey (1977), is an abbreviated histogram, as you can see in the graph below (Figure 2.4). This graph shows the combined sample of 25 DBH activity levels taken from the both the nonpsychotic and the psychotic schizophrenic patients (plotted as dots along a horizontal axis) and a box plot of the same data. The rectangle in the box plot extends from the lower quartile to the upper quartile of the sample data, and is divided at the median. The horizontal lines extend from the ends of this rectangle as far as the extreme data values within a distance equal to the interquartile range. Circles at the ends denote more extreme data values ('outliers').

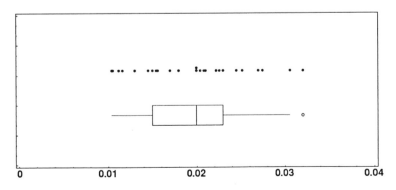

FIGURE 2.4: Box Plot of DBH Activity for 25 Schizophrenics

Note that the median of the 25 DBH levels is 0.0200, the lower quartile is 0.0145, and the upper quartile is 0.0226, so the interquartile range is 0.0081. These values define the central box in the graph. The outliers are defined as data below 0.0064 (0.0145 − 0.0081) or above 0.0307 (0.0226 + 0.0081). In fact there are no data below 0.0064 and only one observation (0.0320) exceeds 0.0307, so there is a single out-lier. Note also that the horizontal lines at each end of the box only extend as far as the data permit, so the left-hand line extends to 0.0104, not to 0.0064, and similarly the right-hand line extends to 0.0306, not to 0.0307.

Figure 2.5 shows box plots of the two samples of DBH levels, from which the assumption that the populations have similar standard deviations may be assessed by comparing the lengths of the boxes. On the basis of this graph this assumption appears to be quite reasonable.

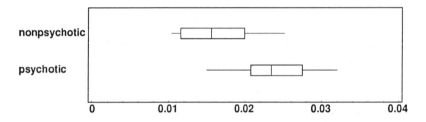

FIGURE 2.5: Box Plots of DBH Activity by Psychotic Status

Figure 2.5 may be compared with Figure 2.3, which shows con-fidence intervals for the population means. The box plots show the distribution of the data, while the confidence intervals show the extent to which the population means could differ. Even though the box plots show considerable overlap in the data from the two samples, the con-fidence intervals do not overlap at all and provide strong evidence that the population means are different.

A p-value is obtainable by using the *two-sample t-test*. In this case the null hypothesis states that the population means are the same, and the t-statistic is obtained by dividing the difference between the sample means by the standard error in Equation (2.5). This gives

$$ t = \frac{\bar{y}_1 - \bar{y}_2}{s\sqrt{\frac{1}{n_1} + \frac{1}{n_2}}} \tag{2.7} $$

where s is given by Equation (2.6). The p-value is then obtained by computing the area in the two tails of the t-distribution with $n_1 + n_2 - 2$ degrees of freedom. In the present case, the difference in means is

0.00783 and the standard error is 0.00199 so the t-statistic is 0.00783 /
0.00199 = 3.94 with 23 degrees of freedom, giving a p-value 0.00066.
This p-value confirms the evidence from the graph of the confidence
intervals.

There is a further assumption underlying the two-sample t-test,
namely, that the data arise from normally distributed populations. You
can assess this assumption to some extent by examining the box plots,
which are very effective in showing skewness. However, normality is
much better assessed by examining a *normal scores plot*, defined as
follows. For a sample of any size n, the normal scores are defined as
the values, in ascending order, that divide the area under the standardised normal curve into $n + 1$ equal components. A normal scores
plot is then obtained by plotting the ordered sample values against the
corresponding normal scores. If the data are sampled from a normal
distribution the points in this plot should tend to follow a straight line.

When assessing the normality assumption for a two-sample t-
test, it is conventional to plot the *residuals* (that is, the observations
adjusted by subtracting their respective sample means) against their
corresponding normal scores, as shown in Figure 2.6.

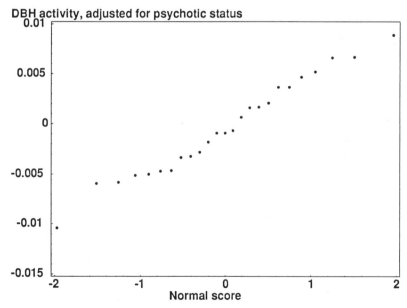

FIGURE 2.6: Normal Scores Plot – 25 Schizophrenics

The points graphed in Figure 2.6 are approximately linear, so
the normality assumption appears to be valid in this case. A less
subjective measure of normality is obtained by computing a p-value

based on a test for normality such as the *Shapiro–Wilk test*, with small *p*-values providing evidence against normality. The *p*-value for this test based on the above data turns out to be 0.78, so the normality assumption is indeed quite reasonable in this case.

The question of what to do when the statistical assumptions are not reasonable is taken up in the next section.

In this section we have considered the comparison of two population means, which is the same as examining the association between a dichotomous determinant and a continuous outcome. The most appropriate statistical methods may be summarised as follows.

First, the data from the two samples should be graphed as box plots. These graphs are useful for showing departures from the statistical assumption that the outcome variable has the same standard deviation in each population, and they also highlight skewness and outliers in the data. Confidence intervals should also be graphed, both for the means and for their difference. Finally, the two-sample *t*-test is used to compute a *p*-value for testing the null hypothesis of equal population means, after checking the statistical assumptions.

5: Paired Data

When investigating the effect of a treatment or exposure on some outcome, the groups being compared need not be independent of each other. In fact there is an advantage in just having one group of subjects, and comparing the effects of the two treatments *on the same subject*. If this can be done, the variation between the subjects is removed from consideration. Studies in which two treatments being compared are applied to the same subject are called *matched* studies.

It is not always feasible to apply different treatments to the same subject. It cannot be done if the effect of a treatment is long-lasting. For example it would not be feasible to use a matched study design to compare the DBH activity levels of psychotic and nonpsychotic patients, because patients do not tend to change from one condition to the other. However, there are many situations where a matched design is possible, and a typical example is now considered.

Brian Everett, Professor of Statistics in the Department of Psychiatry at the University of London, reported data from an experiment in which the body weights of 17 young girls suffering from anorexia were measured before and after they were given a family therapy treatment designed to increase their weight. The data are listed in Table 2.3, where the weights are given in kilograms (see Hand et al (1994), page 229, where the data are expressed in pounds).

You could analyse these data by regarding the pre-treatment

measurements as a sample from an untreated population and the post-treatment measurements as a sample from another population of treated subjects. The methods described in the preceding section could then be used: box plots to display the data, confidence intervals for the means and the difference in means, and a two-sample t-test to decide if the treatment is effective. The box plots are shown in Figure 2.7.

Subject	Before	After
1	38.0	43.2
2	37.8	42.8
3	39.0	41.5
4	37.4	41.7
5	39.3	45.5
6	36.1	34.8
7	34.9	34.8
8	42.7	46.1
9	33.3	43.1
10	36.5	34.1
11	37.0	35.3
12	37.3	43.3
13	35.2	41.2
14	37.9	42.0
15	40.8	42.6
16	39.0	41.6
17	39.6	44.5

TABLE 2.3: Weights of Anorexic Girls given Family Therapy

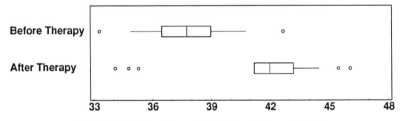

FIGURE 2.7: Box Plots of Weights of Anorexic Girls

The pre- and post-treatment means are 37.75 kg and 41.06 kg, respectively, with standard deviations 2.27 kg and 3.86 kg, so using formulas (2.5) and (2.6) the standard error of the difference in means is 1.08 and a 95% confidence interval is (1.15, 5.47). The two-sample t-test from Equation (2.7) gives $t = 3.31 / 1.08 = 3.06$ with 32 degrees of freedom, corresponding to a p-value 0.005. So you would reject the null hypothesis and conclude that the mean weight gain is most probably between 1.15 kg and 5.47 kg.

A problem with this analysis is that it rests on an assumption that the samples are independent, and this assumption is clearly not valid because the data are taken from the same subjects. One way of solving the problem is to just subtract the pre-treatment measurement from the corresponding post-treatment measurement, thus reducing the data to a single sample of (independent) differences. The standard error of the mean of the differences is given by Equation (2.4), and a confidence interval is obtained in the usual way.

The mean of the sample of weight gains is 3.31 kg, and the standard deviation is 3.26 kg, so Equation (2.4) gives the standard error $3.26/\sqrt{17} = 0.79$, and, using the two standard error rule, a 95% confidence interval for the weight gain is (1.73, 4.89).

The conventional method for testing the null hypothesis that the population means are equal using pair-matched data is called the *paired t-test*, and is based on a *t*-statistic with $n - 1$ degrees of freedom, obtained by dividing the mean of the differences by the standard error of these differences, given by Equation (2.4). This gives

$$t = \frac{\bar{y}_1 - \bar{y}_2}{s/\sqrt{n}} \qquad (2.8)$$

where in this case s is simply the standard deviation of the differences $y_1 - y_2$. For the null hypothesis that the anorexic girls' weight gain is 0, this gives $t = 3.31/0.79 = 4.20$ with 16 degrees of freedom, corresponding to a *p*-value 0.0007.

Since the paired *t*-test avoids the variation between the subjects (by using each subject as their own control) it often gives a more accurate result than the two-sample t-test applied to the same data. Similarly the 95% confidence interval based on the matched analysis is usually shorter than that based on the unmatched analysis. For the anorexic girls the 95% confidence interval based on the matched analysis covers a range of 3.16 kg, compared with 4.32 kg based on the unmatched analysis.

The Need for a Control Group

If a paired *t*-test gives a significant result, does it mean that the treatment is effective? Not necessarily. In the experiment with the anorexic girls, these girls could have gained weight even if they had not been treated. It often happens that patients in clinical trials respond just to the idea of a treatment. So an experiment that involves comparing measurements of some outcome before and after an exposure or treatment does not necessarily prove that an observed change is associated with the exposure. A control group is needed.

In the experiment with the anorexic girls there actually was a control group containing 26 anorexic girls (randomly selected from the same population as the treated group) who were not given any treatment, and their weights at the beginning and end of the observation period are given below in Table 2.4.

The effect of the treatment may now be assessed by comparing the weight gains in the treated group with those in the control group, using the methods for comparing two samples as described in the preceding section. For the control group of anorexic girls the mean weight gain is −0.22 kg (a slight decrease in weight) and the standard deviation is 3.62 kg. For the treated group the mean increase is 3.31 kg with standard deviation 3.26 kg, so the difference in means is 3.53 kg and, using Equation (2.5), the standard error of this observed difference is 1.09 kg. A 95% confidence interval for the weight gain associated with the treatment extends from $3.53 - 2 \times 1.09$ to $3.53 + 2 \times 1.09$ kg, that is, (1.35, 5.71). The two-sample t-statistic is $3.53 / 1.09 = 3.25$ with $26 + 17 - 2 = 41$ degrees of freedom, giving a p-value 0.002.

Subject	Before	After	Subject	Before	After
1	36.6	36.4	14	35.4	36.9
2	40.6	36.3	15	32.0	37.1
3	41.7	39.2	16	35.1	35.1
4	33.6	39.2	17	38.7	38.2
5	35.4	34.5	18	39.0	34.2
6	40.1	35.4	19	38.2	36.1
7	39.6	34.1	20	36.2	33.1
8	34.1	39.3	21	38.8	40.1
9	36.6	33.3	22	38.3	38.4
10	35.6	38.4	23	36.1	36.9
11	35.2	35.1	24	35.2	36.8
12	40.2	36.1	25	32.8	40.0
13	36.9	40.7	26	40.4	35.8

TABLE 2.4: Weights of Anorexic Girls in Control Group

Graphical Displays

For matched studies with continuous outcomes the data may be plotted in two different ways, either as a scatter plot or as a line segment plot. For the study involving the anorexic girls these plots are shown next.

The scatter plots of the two groups are shown first (Figure 2.8). The final measurements are plotted vertically as y-coordinates and the initial measurements are plotted horizontally as x-coordinates.

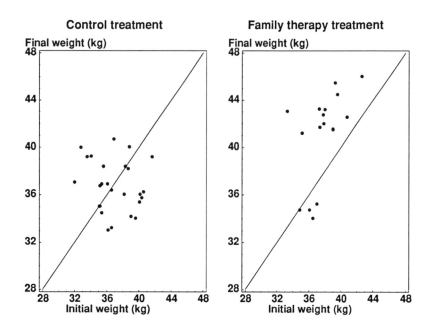

FIGURE 2.8: Scatter Plots of Weights of Anorexic Girls

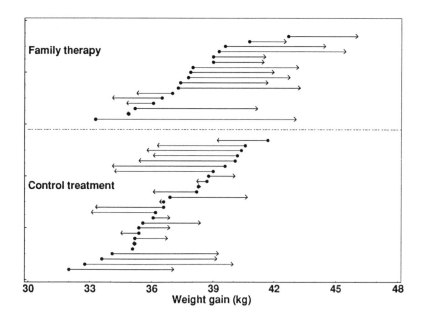

FIGURE 2.9: Parallel Line Plots of Weights of Anorexic Girls

It is useful to draw a line on each scatter plot corresponding to $y = x$, that is, no change, so that points above this line represent increases and points below the line represent decreases.

The scatter plots show that there is no relation between the initial and final weights in the control group, but in the treated group all but four of the points are above the line, with one point on the line and three below it. This means that most of the anorexic girls in the treated group have gained weight, while four have not improved.

A second method for displaying data from a matched study involves plotting each pair of measurements as a horizontal line segment, with these lines ordered vertically according to the values of the initial measurements (see McNeil (1992)), as shown in Figure 2.9.

You can see from Figure 2.9 that most of the line segments in the treated group are displaced to the right, indicating a weight gain. However, in the control group there are about as many lines displaced to the left as to the right, indicating no overall change. Note that the higher initial measurements tend to be displaced to the left and the lower initial measurements tend to be displaced to the right. This effect is called *regression to the mean*, and is explained by the fact that a higher than average measurement is more likely to decrease than to increase, and similarly a lower than average measurement is more likely to increase than to decrease.

Data Transformation

A measure of the severity of a parasitic disease is the concentration of organisms in a specimen sample taken from a subject. In the case of hookworm, a disease common in certain rural areas of Asia, egg concentrations measured from stool samples provide such a measure of severity. Samples taken from 94 infected residents in a fishing community in Southern Thailand in 1993 had egg counts per gram distributed according to the histogram shown in the top panel of Figure 2.10.

The distribution of the egg counts is heavily skewed, and as a result the typical egg count/gram for an infected subject in the population is unstable, varying markedly from sample to sample. In this sample, for example, the mean is 1768 eggs per gram and the median is only 437 eggs per gram.

The distribution is more symmetric if the concentrations are transformed using logarithms, as you can see from the lower panel of Figure 2.10. The mean of the base 10 logarithms of the egg counts is 2.72, while the median is 2.64, only slightly less.

It is also difficult to carry out statistical analysis when samples are skewed, so there is a further advantage in transforming data to remove or reduce skewness. Having done the analysis, the data are easily re-expressed on the original scale if preferred. But in many situations it is best to leave the data in the transformed scale. Many common measurements, such as earthquake intensities (Richter scale), alkaline concentrations (pH), and noise levels (decibels) are naturally expressed on a logarithmic rather than a linear scale.

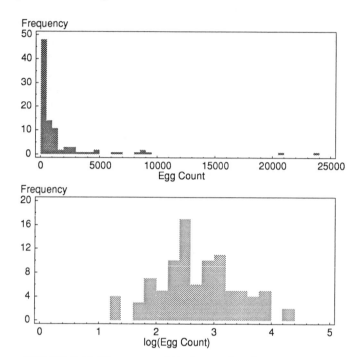

FIGURE 2.10: Histograms of Hookworm Egg Concentrations in Stools

The data in Table 2.5 were reported by Dale et al (1987).

Runner	Before	After
1	4.3	29.6
2	4.6	25.1
3	5.2	15.5
4	5.2	29.6
5	6.6	24.1
6	7.2	37.8
7	8.4	20.2
8	9.0	21.9
9	10.4	14.2
10	14.0	34.6
11	17.8	46.2

TABLE 2.5: Plasma beta Endorphin levels of 11 Fun Runners

These data are plasma beta endorphin concentrations (in pmol/litre) taken from 11 runners before and after they completed a half marathon race. Figure 2.11 shows box plots of the data, both raw and transformed by taking logarithms. The raw data samples have skewed distributions with different spreads before and after the race, while the logarithms are more symmetric with similar spreads. In view of this, you might think that the data should be re-expressed as logarithms prior to statistical analysis.

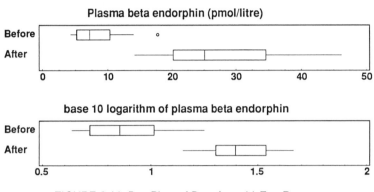

FIGURE 2.11: Box Plots of Data from 11 Fun Runners

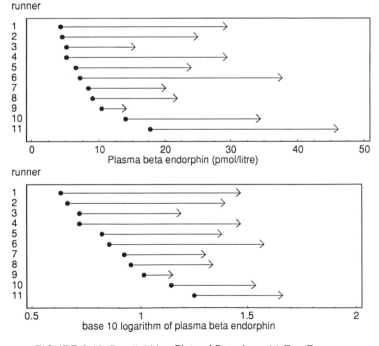

FIGURE 2.12: Parallel Line Plots of Data from 11 Fun Runners

Note, however, that when comparing means from matched samples it is the differences, rather than the data themselves, that are assumed to be sampled from a normal distribution. For this reason Figure 2.12, which shows a parallel line plot of the two samples, provides a better indication of the need for a transformation.

The graphs in Figure 2.12 suggest that the absolute increases in beta endorphin are independent of the initial levels, but proportionately greater at the lower levels, so there is no justification for transforming the data. It would be worthwhile to check that the increases satisfy the normality assumption, and this is left as an exercise.

Further examples of data transformations are given in the next chapter, where we consider situations where the determinant is multi-categorical.

To summarise, in this section you have seen methods for analysing matched pairs of observations on a continuous outcome. These methods include graphing the data using both scatter plots and parallel line plots, computing a confidence interval for the mean of the differences, using a paired t-test to obtain a p-value for testing the null hypothesis that the mean difference in the population is zero, and assessing the normality assumption, both graphically and on the basis of a statistical test. You have also seen how an appropriate data transformation, such as taking logarithms, can improve the validity of the statistical assumptions.

Matched designs also arise when outcomes are categorical. However, the methods for analysing such data are a little beyond the level of this introductory chapter, and are considered in detail in Chapter 8.

Summary

This chapter has summarised the basic statistical methods used in the analysis of data from comparative studies, assuming a single, dichotomous, determinant variable and a single outcome variable, which could be categorical or continuous. These methods include tabular and graphical presentation of data and confidence intervals, hypothesis tests, methods for assessing statistical assumptions, and data transformation to reduce skewness and heterogeneity.

Two-by-two Tables

The results from a study investigating an association between a dichotomous determinant and a dichotomous outcome may be shown in a two-by-two contingency table. Using data from a randomised trial comparing the effects of home and hospital care on the survival of heart patients, you saw how the hypothesis of no association may be tested using Pearson's chi-squared test, and confidence intervals may be obtained for proportions and odds ratios.

Multiple Outcomes

Pearson's test extends to data involving multicategorical outcomes, where it is natural to use odds ratios to analyse the association. These methods were illustrated with an example involving the comparison of different toxicity grades in two treatment arms of a clinical trial involving patients with breast cancer.

Continuous Outcomes

Box plots are effective graphs for displaying continuously varying data, and graphs of confidence intervals for population means provide an effective method for assessing the extent of the difference in these means. Provided the populations being compared are normal with a common standard deviation, the two-sample t-test may be used to test the null hypothesis, and the normality assumption may be assessed by plotting residuals against normal scores. Dopamine activity levels taken from two groups of hospitalised schizophrenic patients were used to illustrate these methods.

Paired Data

Using weights of anorexic girls before and after a treatment as an illustration, you saw how different methods are needed to handle matched data. Graphical methods include scatter plots and parallel line plots of the data and the paired t-test for testing the null hypothesis. In this section the concept of data transformation was also introduced. Measurements with skewed distributions are difficult to analyse for a number of reasons: means and medians can differ markedly, different subgroups have different spreads, and statistical methods based on normality assumptions are not valid. Taking a transformation such as the logarithm often overcomes these difficulties.

Exercises

Exercise 2.1: The study of home-versus-hospital management for heart disease subjects whose results are shown in Table 2.1 did not give a conclusive result. Suppose you had carried out a similar study with ten times as many subjects (1320 in each treatment arm) and you obtained the same sample proportions. Would your study be conclusive? Answer the question by repeating the analysis given in Section 2.

Exercise 2.2: The following contingency table is taken from a clinical trial reported by Ezdinli et al (1976) and further discussed by Pocock (1983, page 191), in which two treatments for 273 patients with lymphomas were compared.

Test the hypothesis that there is no association between treatment and outcome, and graph the odds ratios comparing the two treatments at each outcome level, treating 'complete response' as the referent outcome.

Outcome	Treatment	
	BP	CP
Complete Response	26	31
Partial Response	51	59
No Change	21	11
Progression	40	34

Exercise 2.3: The following data are steady-state haemoglobin levels for patients with different types of sickle cell disease, reported by Anionwu et al (1981).

Graph these data using box plots, and compare the population means in the three groups using three separate two-sample *t*-tests.

HB SS	HB S/-thalassaemia	HB SC
7.2	8.1	10.7
7.7	9.2	11.3
8.0	10.0	11.5
8.1	10.4	11.6
8.3	10.6	11.7
8.4	10.9	11.8
8.4	11.1	12.0
8.5	11.9	12.1
8.6	12.0	12.3
8.7	12.1	12.6
9.1		12.6
9.1		13.3
9.1		13.3
9.8		13.8
10.1		13.9
10.3		

Exercise 2.4: The study of anorexic girls considered on page 47 involved a third group a subjects who were given 'cognitive behav-

ioural therapy' treatment, and the complete set of data is as follows. (Note that all the measurements are expressed as the original units, pounds rather than kilograms.)

Does the cognitive behavioural therapy treatment work? Answer this question by (a) examining the weight gains in the group given the behavioural therapy treatment, and (b) comparing these weight gains with those in the control group.

Cognitive Behavioural		Control		Family Therapy	
before	after	before	after	before	after
80.5	82.2	80.7	80.2	83.8	95.2
84.9	85.6	89.4	80.1	83.3	94.3
81.5	81.4	91.8	86.4	86.0	91.5
82.6	81.9	74.0	86.3	82.5	91.9
79.9	76.4	78.1	76.1	86.7	100.3
88.7	103.6	88.3	78.1	79.6	76.7
94.9	98.4	87.3	75.1	76.9	76.8
76.3	93.4	75.1	86.7	94.2	101.6
81.0	73.4	80.6	73.5	73.4	94.9
80.5	82.1	78.4	84.6	80.5	75.2
85.0	96.7	77.6	77.4	81.6	77.8
89.2	95.3	88.7	79.5	82.1	95.5
81.3	82.4	81.3	89.6	77.6	90.7
76.5	72.5	78.1	81.4	83.5	92.5
70.0	90.9	70.5	81.8	89.9	93.8
80.4	71.3	77.3	77.3	86.0	91.7
83.3	85.4	85.2	84.2	87.3	98.0
83.0	81.6	86.0	75.4		
87.7	89.1	84.1	79.5		
84.2	83.9	79.7	73.0		
86.4	82.7	85.5	88.3		
76.5	75.7	84.4	84.7		
80.2	82.6	79.6	81.4		
87.8	100.4	77.5	81.2		
83.3	85.2	72.3	88.2		
79.7	83.6	89.0	78.8		
84.5	84.6				
80.8	96.2				
87.4	86.7				

Exercise 2.5: The following table gives blood pressure measurements in mm Hg taken from 15 hypertensive patients before and two hours after treatment with captopril. The data were reported by MacGregor et al (1979) and are also discussed by Cox and Snell (1981).

(a) Graph the data, and provide a suitable comment to go with the graph.

(b) Test that the decreases are normally distributed, and give 95% confidence intervals for the mean decreases in both systolic and diastolic blood pressure.

Patient number	Systolic before	Systolic after	Diastolic before	Diastolic after
1	210	201	130	125
2	169	165	122	121
3	187	166	124	121
4	160	157	104	106
5	167	147	112	101
6	176	145	101	85
7	185	168	121	98
8	206	180	124	105
9	173	147	115	103
10	146	136	102	98
11	174	151	98	90
12	201	168	119	98
13	198	179	106	110
14	148	129	107	103
15	154	131	100	82

References

Anionwu, E., D. Watford, M. Brozovic & B. Kirkwood (1981): Sickle cell disease in a British urban community, *British Medical Journal*, **282**, pages 283–286.

Cox, D.R. & E.J. Snell (1981): *Applied Statistics*. Chapman & Hall. London.

Dale, G., J.A. Fleetwood, A. Weddell, R.D. Ellis & J.R.C. Sainsbury (1987): Beta endorphin: a factor in 'fun run' collapse? *British Medical Journal*, **294**, page 1004.

Ezdinli, E. et al (1976): Comparison of intensive versus moderate chemotherapy of lymphocytic lymphomas: a progress report, *Cancer*, **38**, pages 1060–1068.

Hampton J.R. & J.D. Hill (1978): A randomized trial of home-versus-hospital management for patients with suspected myocardial infarction, *The Lancet*, 22 April 1978, pages 837–841.

Hand, D.J., F. Daly, A.D. Lunn, K.J. McConway & E. Ostrowski (1993): *A Handbook of Small Data Sets*. Chapman & Hall. London.

MacGregor, G.A., N.D. Markandu, J.E. Roulston, & J.C. Jones (1979): Essential hypertension: effect of an oral inhibitor of angiotensin-converting enzyme, *British Medical Journal*, **2**, pages 1106–1109.

McNeil, D. (1992): On graphing paired data, *the American Statistician*, **46**, pages 307–311.

Pocock, S. (1983): *Clinical Trials*. John Wiley & Sons. Chichester.

Sternberg, D.E., D.P. Van Kammen & W.E. Bunney (1982): Schizophrenia: dopamine *b*-hydroxylase activity and treatment response, *Science* **216**, pages 1423–1425.

Tukey, J.W. (1977): *Exploratory Data Analysis*. McGraw-Hill. Reading: MA.

3

STATISTICAL METHODS II

1: Introduction

Chapter 2 provided an outline of the basic statistical methods for analysing data from the simplest comparative studies, that is, studies involving two samples and no covariates. In the present chapter these methods are extended to situations where the determinant is either multicategorical or continuous.

We begin with a method for the analysis of general contingency tables with more than two rows and columns, where Pearson's chi-squared test is still appropriate for testing the hypothesis of no association between the determinant and outcome, and graphs of odds ratios are very effective for showing the pattern of the association. Next we show that one-way analysis of variance is the natural extension of the two-sample t-test for comparing means from several populations, and leads to the important problem of multiple comparison. Turning to the analysis of matched designs, the two-way analysis of variance procedure is shown to be the extension of the paired t-test.

Next we give an outline of the linear regression model as a method for analysis when both the determinant and outcome variables are defined on a continuous range. This regression model provides a basis for further methods of interest to epidemiologists, including the methods for handling dichotomous and multicategorical outcomes with a continuous determinant developed in the chapters to follow.

Chapter 3 concludes by showing how a generalisation of the simple linear regression model can provide further insights into the problem of comparing means from several populations.

As in Chapter 2, graphical methods play an essential role, both for displaying data and for assessing the statistical assumptions that underlie the methods, and the role of data transformations is further explored. The methods are illustrated with a variety of examples taken from the scientific literature. These include a study of snoring as a risk factor for heart disease, an examination of the association between eye colour and hair colour in schoolchildren, a comparison of the survival times of patients with different cancers, an analysis of scores given by different judges to contestants in a synchronised swimming contest, an investigation of the relationship between the level of an environmental industrial contaminant and the shell thicknesses of pelican eggs, and a comparison of silver contents of different samples of Byzantine coins.

2: General Contingency Tables

In Section 3 of Chapter 2 an example was given involving an outcome classified into five categories (toxicity grade) and a dichotomous determinant (one or other of two treatments). The methods described in that section apply equally well to situations involving a dichotomous outcome and a multicategorical determinant, simply by reversing the roles of determinant and outcome. As an illustration, Table 3.1 shows data from a cross-sectional study of snoring behaviour as a risk factor for heart disease, reported by Norton and Dunn (1985).

| | | *Snoring behaviour* | | | | |
		never	occasional	frequent	always	total
Heart	yes	24	35	21	30	110
disease	no	1355	603	192	224	2374
	total	1379	638	213	254	2484

TABLE 3.1: Results from a Snoring Survey

The null hypothesis of no association may be tested by computing a chi-squared statistic which measures the discrepancy between the observed counts and expected counts based on the null hypothesis, using the method given in Section 3 of Chapter 2. This procedure gives the chi-squared statistic 72.78 with degrees of freedom 3 (one less than the number of categories into which the determinant is classified). The p-value is the area of the upper tail of the chi-squared distribution beyond the observed statistic, and is very close to 0 in this case, so the null hypothesis should be rejected. The conclusion is that there is an association between snoring and heart disease.

To calculate and graph the odds ratios, let us choose non-snorers as the referent category. The odds of heart disease is 35/603 for the occasional snorers and 24/1355 for non-snorers, so the odds ratio of occasional to non-snorers is (35/603) / (24/1355) = 3.28, with natural logarithm 1.18. The corresponding standard error is thus $\sqrt{(1/35+1/24+1/603+1/1355)}$ = 0.27, so a 95% confidence interval for the odds ratio is ($e^{0.64}$, $e^{1.72}$), or (1.9, 5.58). Similarly, the odds ratio for comparing the risks between the frequent snorers and the non-snorers is (21/192) / (24/1355) = 6.18, whose natural logarithm has standard error 0.31, giving the 95% confidence interval (3.32, 11.47) for the odds ratio. In the same way the 95% confidence interval for the odds ratio for heart disease for those who always snore compared with the non-snorers is computed to be (4.31, 13.33). These intervals are displayed in the upper panel of Figure 3.1 below.

Snoring comparison

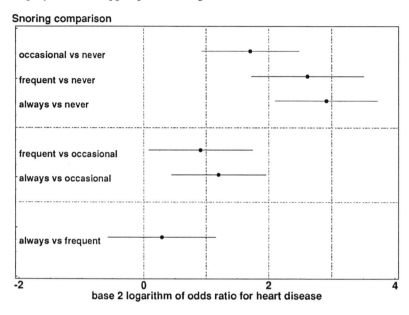

base 2 logarithm of odds ratio for heart disease

FIGURE 3.1: Association between Snoring and Heart Disease

The confidence intervals are graphed on a base 2 logarithmic scale, so that each unit on the scale corresponds to a doubling of the odds. Thus you can see easily from the graph that the chronic snorers have odds of getting heart disease between just over four (2^2) and just under 16 (2^4) times the odds of non-snorers.

The middle panel of the graph shows the confidence intervals for the comparisons between most frequent snorers and occasional snorers, while the bottom panel compares the two groups of frequent snorers. Since the last confidence interval includes 0, the two groups of frequent snorers have similar risks of heart disease. We conclude that there are three groups with different risks: the non-snorers, the occasional snorers, and those who snore frequently or all the time.

When the result of a multiple comparison is that two or more samples are statistically indistinguishable, it may be reasonable to combine them. Thus we arrive at the graph shown in Figure 3.2, which confirms the conclusion reached from the preceding analysis, that is, that there are three distinct risk groups.

Snoring comparison

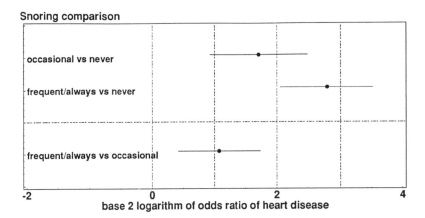

FIGURE 3.2: Simplified Association between Snoring and Heart Disease

It should be noted that even if samples may be combined using statistical considerations, it may not make clinical sense to do so. However, in the case of the snorers there seems no clinical reason not to combine the two most frequent snoring groups.

The situation is more complicated when both the determinant and the outcome are multicategorical, but the methods still apply. As an illustration, consider the following data (shown below in Table 3.2) relating hair and eye colour from children in Scotland analysed by Goodman (1981). In this case we do not have a determinant and an outcome (in fact both factors are outcomes of a genetic determinant) but it is still of interest to investigate a possible association between the two variables. For purely statistical reasons we will take eye colour as the 'outcome' and hair colour as the 'determinant'.

| | | *Hair colour* | | | | |
		fair	red	medium	dark	black
	blue	326	38	241	110	3
Eye	light	688	116	584	188	4
colour	medium	343	84	909	412	26
	dark	98	48	403	681	85

TABLE 3.2: Hair and Eye Colour for 5387 Scottish Children

Pearson's chi-squared test may again be applied to test the null hypothesis of no association. As before, the test statistic is computed by comparing the observed counts with expected counts obtained by dividing the product of the marginal totals by the overall total, and the residuals are squared, divided by the expected counts, and summed. The result is compared with a chi-squared distribution whose degrees of freedom is the product $(r-1)(c-1)$, where r is the number of outcome categories and c is the number of categories for the determinant. As before, the p-value is the area in the upper tail of the distribution.

For these data the test statistic turns out to be 1240.0 and the number of degrees of freedom is $(5-1)(4-1) = 12$, so the association is clearly significant, giving a p-value very close to 0.

If a referent category is chosen for each variable, odds ratios may be calculated for each combination of the determinant and outcome categories in the usual way. The following graph (Figure 3.3) shows 95% confidence intervals for the various odds ratios, taking fair hair and blue eyes as the referent categories.

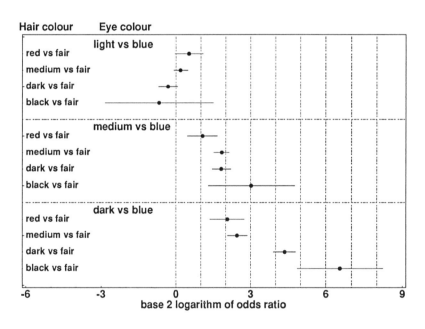

FIGURE 3.3: Hair & Eye Colour Comparison for Scottish Children

You can see from the graph of the confidence intervals that there is not much difference between light-coloured and blue eyes as far as the association with hair colour is concerned. However, children with red or darker hair colour have different eye colour distributions, and are more likely to have darker eyes than those with fair hair.

The remaining sections of this chapter deal with continuous outcomes. Sections 3 and 4 generalise the methods described in Sections 4 and 5, and in Sections 5 and 6 the linear regression model is introduced as a general procedure for analysing continuous outcome data.

3: One-way Analysis of Variance

In this section we consider methods for the analysis of data in which the outcome is continuous and the determinant is categorical. This leads to a procedure called 'the analysis of variance' (anova).

Table 3.3 shows survival times in days for 64 patients with cancers of different organs, reported by Cameron and Pauling (1978).

Stomach	Bronchus	Colon	Ovary	Breast
124	81	248	1234	1235
42	461	377	89	24
25	20	189	201	1581
45	450	1843	356	1166
412	246	180	2970	40
51	166	537	456	727
1112	63	519		3808
46	64	455		791
103	155	406		1804
876	859	365		3460
146	151	942		719
340	166	776		
396	37	372		
	223	163		
	138	101		
	72	20		
	245	283		

TABLE 3.3: Survival Times in Days of Cancer Patients

A problem of interest here is to compare the survival times for the different organs, to see how the type of cancer affects a patient's survival prospects. The data may be displayed using box plots, as shown in the next graph (Figure 3.4). The distributions are clearly skewed, which complicates the comparison of the survival times for the different organs, and we will address this issue before completing this section.

The first question to ask when comparing the outcomes in several groups is 'could the samples have arisen from the same population?' In this case the null hypothesis may be stated as

$$H_0: \mu_1 = \mu_2 = \ldots = \mu_c$$

where c is the number of groups and μ_j is the population mean corresponding to group j.

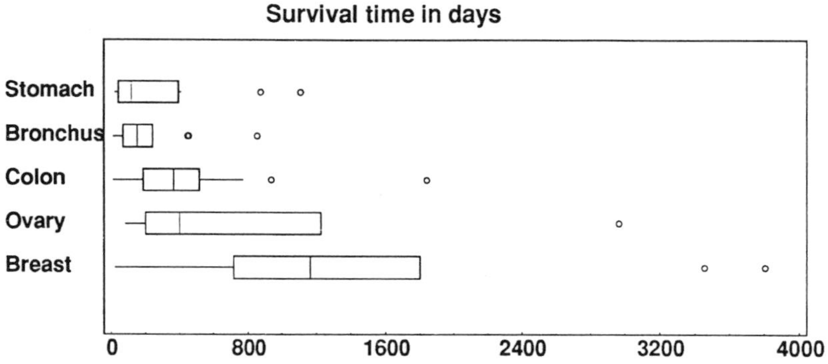

Survival time in days

FIGURE 3.4: Box Plots of Survival Times of Cancer Patients

This null hypothesis may be tested by computing a statistic called the *F-statistic* and comparing it with an appropriate distribution to get a *p*-value. Suppose that there are n_j observations in sample j, denoted by y_{ij} for $i = 1, 2, ..., n_j$. The *F*-statistic is defined as

$$F = \frac{(S_0 - S_1)/(c-1)}{S_1/(n-c)} \qquad (3.1)$$

where

$$S_0 = \sum_{j=1}^{c} \sum_{i=1}^{n_j} (y_{ij} - \bar{y})^2, \quad S_1 = \sum_{j=1}^{c} \sum_{i=1}^{n_j} (y_{ij} - \bar{y}_j)^2$$

and

$$\bar{y}_j = \frac{1}{n_j} \sum_{i=1}^{n_j} y_{ij}, \quad \bar{y} = \frac{1}{n} \sum_{j=1}^{c} \sum_{i=1}^{n_j} y_{ij}, \quad n = \sum_{j=1}^{c} n_j$$

Note that S_0 is the sum of squares of the data after subtracting their overall mean, while S_1 is the sum of squares of the residuals obtained by subtracting each sample mean. If the population means are the same the numerator and the denominator in the *F*-statistic are independent estimates of the square of the population standard deviation (assumed the same for each population) and the *p*-value is the area in the tail of the *F*-distribution with $c-1$ and $n-c$ degrees of freedom.

For the survival times of the 64 cancer patients, these formulas give $S_0 = 37\,983\,904$ and $S_1 = 26\,448\,146$, so

$$F = \frac{(37983904 - 26448146)/(5 - 1)}{26448146/(64 - 5)} = 6.43$$

and with 4 and 59 degrees of freedom the area in the tail of the F-distribution beyond this value is 0.0002, indicating that the population means are different. However this procedure, like the two-sample t-test, is based on an assumption that the populations have a common standard deviation, and the box plots in Figure 3.4 clearly show that this assumption is not met.

Figure 3.5 shows box plots of the survival times after transforming the data by taking base 2 logarithms. (Again the logarithms are taken to base 2 because each unit increase in the transformed scale then corresponds to a doubling of the survival time, facilitating interpretation of the transformed data.) You can see that the spreads of these samples in Figure 3.5 are much more even than they are in Figure 3.4.

base 2 logarithm of survival time in days

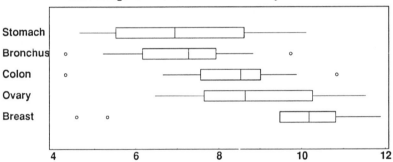

FIGURE 3.5: Log–transformed Survival Times of Cancer Patients

Repeating the one-way anova procedure with the transformed data, you get $S_0 = 226.36$, $S_1 = 175.40$ and $F = 4.29$, giving the p-value 0.004, confirming that the survival times are different.

The F-test also requires the further assumption that the adjusted data (that is, the data adjusted by subtracting the population means from their respective samples) should have arisen from a normal distribution. This assumption may be checked by graphing the residuals against normal scores, as shown in Figure 3.6.

The normal scores plot shows a rough linear trend, suggesting that the normality assumption might be reasonable for these data. More objectively, the Shapiro–Wilk test gives a p-value 0.083, which is not small enough to reject the normality assumption if the conventional 0.05 criterion for statistical significance is used.

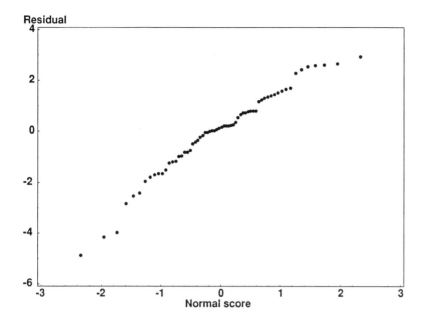

FIGURE 3.6: Normal Scores Plot: Cancer Patients' Survival

Figure 3.7 shows the confidence intervals for the means. The standard errors used to compute these confidence intervals are based on an estimate of the common standard deviation given by the formula

$$s = \sqrt{\frac{S_1}{n - c}} \qquad (3.2)$$

which generalises Equation (2.6). Using Equation (2.4), an estimate of the standard error for the mean in sample j is $s/\sqrt{n_j}$.

You can see from this graph that there is no evidence of a difference in survival for stomach and bronchus cancers, nor is there a statistically significant difference in survival between the breast and ovarian cancer patients.

Figure 3.8 shows the confidence intervals after aggregating the data into just three groups by combining (a) the stomach and bronchus cancers and (b) the breast and ovarian cancers. There is little doubt that the patients with breast and ovarian cancers survive longer than those with cancers of the stomach and bronchus, with patients having cancer of the colon having intermediate survival times. However, the extent to which the data pooling is clinically relevant needs to be considered.

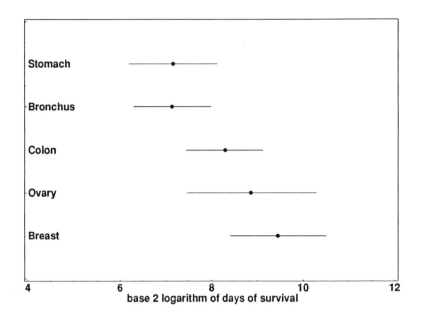

FIGURE 3.7: Confidence Intervals of Survival Means

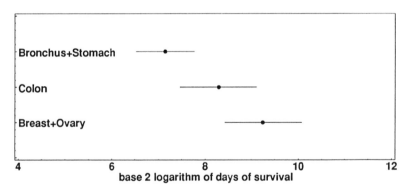

FIGURE 3.8: Confidence Intervals after Pooling Survival Data

In Section 6 you will see how linear regression analysis may be used to cast further light on the question of combining groups. It is important to take both statistical and clinical considerations into account when considering this question.

4: Two-way Analysis of Variance

The one-way anova method is an extension of the method for comparing two means using the two-sample t-test. In the same way, the method for comparing two means based on pair-matched data (using the paired t-test) may be generalised to the comparison of several means. This leads to a method called *two-way anova*, which is considered in this section.

Table 3.6 shows the total scores assigned by each of five judges to each of 40 competitors in a synchronised swimming event at the 1986 U.S. National Olympic Festival in Houston, Texas, as reported by Fligner and Verducci (1988).

Contestant	Judge 1	2	3	4	5	Contestant	Judge 1	2	3	4	5
1	33.1	32.0	31.2	31.2	31.4	21	27.3	28.1	28.4	27.5	26.4
2	26.2	29.2	28.4	27.3	25.3	22	29.5	28.1	27.3	28.4	26.4
3	31.2	30.1	30.1	31.2	29.2	23	28.4	29.5	28.4	28.6	27.5
4	27.0	27.9	27.3	24.7	28.1	24	31.2	29.5	29.2	31.2	27.3
5	28.4	25.3	25.6	26.7	26.2	25	30.1	31.2	28.1	31.2	29.2
6	28.1	28.1	28.1	32.0	28.4	26	31.2	31.2	31.2	31.2	30.3
7	27.0	28.1	28.1	28.1	27.0	27	26.2	28.1	26.2	25.9	26.2
8	25.1	27.3	26.2	27.5	27.3	28	27.3	27.3	27.0	28.1	28.1
9	31.2	29.2	31.2	32.0	30.1	29	29.2	26.4	27.3	27.3	27.3
10	30.1	30.1	28.1	28.6	30.1	30	29.5	27.3	29.2	28.4	28.1
11	29.0	28.1	29.2	29.0	27.0	31	28.1	27.3	29.2	28.1	29.2
12	27.0	27.0	27.3	26.4	25.3	32	31.2	31.2	31.2	31.2	28.4
13	31.2	33.1	31.2	30.3	29.2	33	28.1	27.3	27.3	28.4	28.4
14	32.3	31.2	32.3	31.2	31.2	34	24.0	28.1	26.4	25.1	25.3
15	29.5	28.4	30.3	30.3	28.4	35	27.0	29.0	27.3	26.4	28.1
16	29.2	29.2	29.2	30.9	28.1	36	27.5	27.5	24.5	25.6	25.3
17	32.3	31.2	29.2	29.5	31.2	37	27.3	29.5	26.2	27.5	28.1
18	27.3	30.1	29.2	29.2	29.2	38	31.2	30.1	27.3	30.1	29.2
19	26.4	27.3	27.3	28.1	26.4	39	27.0	27.5	27.3	27.0	27.3
20	27.3	26.7	26.4	26.4	26.4	40	31.2	29.5	30.1	28.4	28.4

TABLE 3.4: Scores awarded to Synchronised Swimming Contestants

A question of interest here is to see if there are differences between the judges' scores. As in the one-way anova method, the data may be displayed using box plots, as shown in the Figure 3.9.

You can see from the box plots that the judges are reasonably consistent, and that their scores have similar spreads. The one-way anova procedure for comparing the means gives sums of squares $S_0 = 704.32$ for the mean-corrected data and $S_1 = 685.23$ for the residuals. The F-statistic is thus $((704.32-685.23)/4) / (685.23/195) = 1.36$ and the p-value is 0.25, suggesting that the null hypothesis that the judges give the same scores as each other should not be rejected.

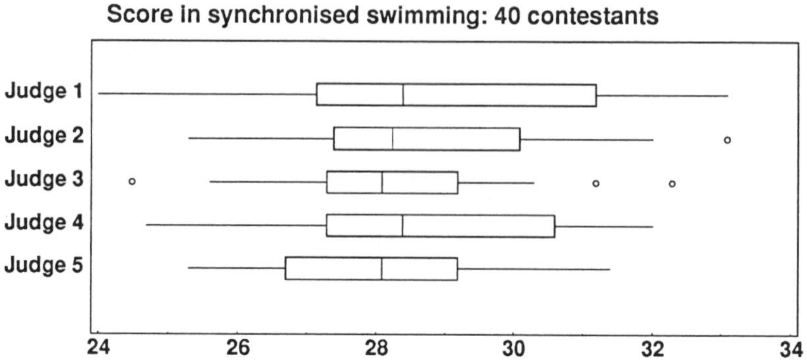

Score in synchronised swimming: 40 contestants

FIGURE 3.9: Box Plots of Synchronised Swimming Scores

The one-way anova fails to take account of the fact that there are 40 contestants each having five repeated measurements taken from them, rather than five samples each containing 40 different contestants. In contrast the two-way anova method assumes that there are matched sets of data, and removes a common effect from each set before comparing the samples. Since the sets (the 40 contestants in this case) are listed as 40 rows of data, this involves subtracting row effects from the data.

To see how one-way anova works, first note that the overall mean of all the scores is 28.53. The mean score for the first contestant is given by $(33.1 + 32.0 + 31.2 + 31.2 + 31.4)/5 = 31.78$, which is 3.25 units above the overall mean. Similarly, for the second contestant the mean score is given by $(26.2 + 29.2 + 28.4 + 27.3 + 25.3) = 27.28$, and this is 1.25 units less than the overall mean. Continuing in this way, you may compute a relative effect for each of the 40 contestants. These effects are both positive and negative, and indicate the relative performances of the contestants. A contestant with a positive effect is above average, while one with a negative effect is below average. For the first four contestants, these relative effects are 3.25, −1.25, 1.83, and −1.53.

Now to compare the judges after adjusting for the effects of the individual differences between the contestants, the contestant effects need to be subtracted from the data in each row of the table of scores. Thus the adjusted scores for the first four contestants are as follows.

Figure 3.10 shows box plots of these adjusted scores. You can see from Figures 3.9 and 3.10 that the spreads of the adjusted scores are much smaller than those of the raw scores. Also the spreads corresponding to the different judges are very similar. Moreover, it now appears that the scores of Judge 5, and possibly those of Judge 3, are lower than those of the other judges.

| | | | | Judge | | |
Contestant	Adjustment	1	2	3	4	5
1	3.25	29.85	28.75	27.95	27.95	28.15
2	−1.25	27.45	30.45	29.65	28.55	26.55
3	1.83	29.37	28.27	28.27	29.37	27.37
4	−1.53	28.53	29.43	28.83	26.23	29.63

Adjusted score: 40 contestants

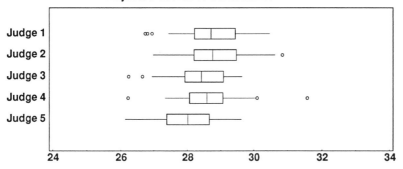

FIGURE 3.10: Adjusted Synchronised Swimming Scores

Having taken into account the differences between the contestants, the judges could be compared by applying the one way anova method to the adjusted scores. However, this method does not give the correct p-value, because the one-way anova procedure requires that the samples be independent of each other. If the data array has r rows and no missing observations (giving $n = r \times c$ observations altogether), a correct p-value is based on an F-statistic defined as

$$F = \frac{(S_2 - S_{12})/(c-1)}{S_{12}/(n-c-r+1)} \qquad (3.3)$$

where

$$S_2 = \sum_{j=1}^{c} \sum_{i=1}^{r} (y_{ij} - \bar{y}_i)^2, \quad S_{12} = \sum_{j=1}^{c} \sum_{i=1}^{r} (y_{ij} - \bar{y}_i - \bar{y}_j + \bar{y})^2$$

and

$$\bar{y}_i = \frac{1}{c} \sum_{j=1}^{c} y_{ij}, \quad \bar{y}_j = \frac{1}{r} \sum_{i=1}^{r} y_{ij}, \quad \bar{y} = \frac{1}{rc} \sum_{j=1}^{c} \sum_{i=1}^{r} y_{ij}$$

The p-value is the area in the tail of the F-distribution which has $c - 1$ and $n - c - r + 1$ degrees of freedom. Note that S_2 is the sum of the squares of the data after adjusting for row effects, S_1, used in Equation (3.1), is the sum of squares after adjusting for column effects, and S_{12} is the sum of squares after adjusting for both row effects and column effects.

For the synchronised swimming scores the F-statistic is $(19.08/4)/(163.96/156) = 4.54$, giving the p-value 0.0017, so the null hypothesis is not tenable. Figure 3.11 shows confidence intervals for the mean scores corresponding to the five judges. The method for computing these confidence intervals is similar to that used in the one-way anova procedure. In this case the formula for the common stand-ard deviation is

$$s = \sqrt{\frac{S_{12}}{n - c - r + 1}} \qquad (3.4)$$

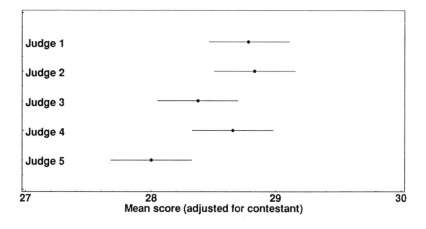

FIGURE 3.11: Confidence Intervals of Judges' mean Scores

From the confidence intervals it appears that Judge 5 gives lower scores than the others, and judges 1, 2, and 4 do not give significantly different scores from each other. Judge 3 may also be similar to judges 1, 2, and 4. So the multiple comparison of the judges' scores could possibly be summarised as follows.

Judge 5 < Judge 3 = Judge 4 = Judge 1 = Judge 2

Having decided that Judge 5 gives lower scores than the others, you could test that the others are equal by repeating the two-way anova procedure after omitting the data for Judge 5. This is left as an exercise.

Statistical Assumptions

One-way anova is based on two assumptions: (1) the standard deviations of the populations are equal, and (2) the populations are normally distributed. The two-way anova method also makes these two assumptions, but the assumptions apply to the data after adjusting for row effects, that is, it is assumed that the population standard deviations of the adjusted populations are all the same and each adjusted population is normally distributed.

As in the case of one-way anova, you can assess the statistical assumptions by first looking at the box plots of the adjusted data. For the judges' scores, the spreads are similar, so the first assumption is reasonable. Turning to the second assumption, you need to examine the residuals from the analysis, which are obtained by subtracting the column means from the adjusted data. As usual, the normality assumption may be assessed graphically by plotting these residuals against their normal scores and seeing if these data could be linear, and the assumption may be tested using the Shapiro–Wilk test. This is left as an exercise.

When these statistical assumptions are violated it may be possible to transform the data, so that the assumptions are reasonable for the transformed data, and it is then feasible to calculate the p-value on the basis of the transformed data.

Anova Plots

A useful graphical display of data analysed using two-way anova is an *anova plot* (see Winer (1971)). In this graph the responses for each column are plotted against the row index, forming a separate panel for each column of the data array.

Figure 3.12 shows such a plot for the synchronised swimming scores, where the contestants are sorted according to their total scores. The continuous lines on the graph correspond to the fitted values obtained by adding the row and column effects, and thus provide a basis for comparing the variation between the samples with the variation within these samples, as well as highlighting extreme observations.

It is clear from the anova plot that the differences between the contestants dominates the differences between the judges. Looking at the lowest score, you can also see that this corresponds to Judge 1 and the lowest scoring contestant (actually Contestant 34), and is approximately 2 units below the curve, that is, it is about 2 units lower than the value given by the fitted model.

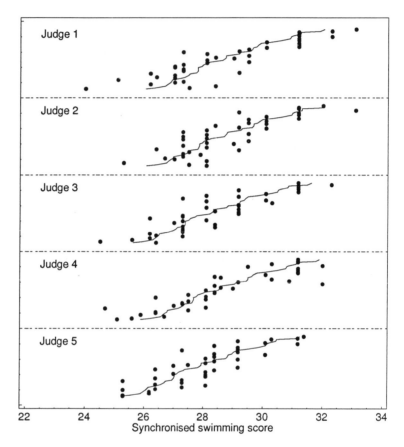

FIGURE 3.12: Anova Plot of Synchronised Swimming Scores

Anova plots may show the data sorted by either rows or columns, and it is often useful to show both graphs. Figure 3.13 shows an anova plot of the synchronised swimming scores with a separate panel for each contestant.

This plot again shows that the variation between contestants is much larger than the variation between judges, and it shows the outliers more clearly than the first anova plot. For example you can see that the largest residual (approximately 3 units) corresponds to Contestant 6 and Judge 4. This plot also highlights the fact that for some reason a lot of scores are clumped at the value 31.2.

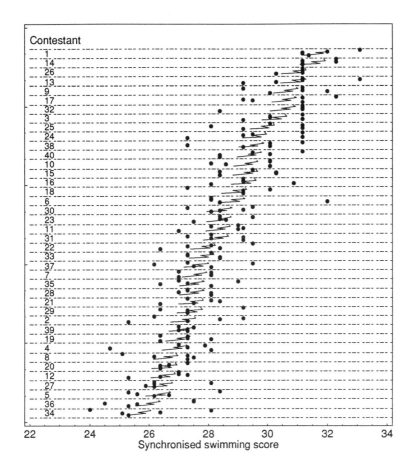

FIGURE 3.13: Alternative Anova Plot of Synchronised Swimming Scores

5: Simple Regression

We have discussed examples in which the outcome variable is dichotomous, multicategorical, or continuous, but in all of these situations the determinant is categorical. In this section we show how to analyse data from studies in which both the determinant and the outcome are continuous variables.

First we show how a scatter plot may be used to display the data, and then we show how the slope of a fitted straight line is used to represent the association between the determinant and the outcome. The usual statistical methods − confidence intervals, p-values, and example from environmental science is used to illustrate these concepts.

Table 3.5 lists some data reported by Risebrough (1972). These are concentrations in ppm of polychlorinated biphenyl (PCB, an industrial pollutant) and shell thickness in mm, measured from 65 Anacapa pelican eggs.

Conc.	Thick.	Conc.	Thick.	Conc.	Thick.	Conc.	Thick.
452	0.14	184	0.19	115	0.20	315	0.20
139	0.21	177	0.22	214	0.22	356	0.22
166	0.23	246	0.23	177	0.23	289	0.23
175	0.24	296	0.25	205	0.25	324	0.26
260	0.26	188	0.26	208	0.26	109	0.27
204	0.28	89	0.28	320	0.28	265	0.29
138	0.29	198	0.29	191	0.29	193	0.29
316	0.29	122	0.30	305	0.30	203	0.30
396	0.30	250	0.30	230	0.30	214	0.30
46	0.31	256	0.31	204	0.32	150	0.34
218	0.34	261	0.34	143	0.35	229	0.35
173	0.36	132	0.36	175	0.36	236	0.37
220	0.37	212	0.37	119	0.39	144	0.39
147	0.39	171	0.40	216	0.41	232	0.41
216	0.42	164	0.42	185	0.42	87	0.44
216	0.46	199	0.46	236	0.47	237	0.49
206	0.49						

TABLE 3.5: PCB Concentration and Shell Thickness of 65 Pelican Eggs

The shell thickness may be regarded as an outcome measurement and the PCB concentration as a determinant. These data may be displayed as a scatter plot in which each subject is denoted by a point with the determinant represented by the distance on a horizontal axis and the outcome as the distance on the vertical axis, as shown in Figure 3.14. However the scatter plot looks rather random and it is difficult to tell if there is any association between the PCB concentration and the shell thickness.

Consider the application of various statistical methods to these data. Suppose we make each variable dichotomous, by grouping the PCB concentrations into 'below 200 ppm' and '200 ppm or above', and by grouping the thicknesses into 'less than 0.3 mm' and '0.3 mm or more'. This gives a 2-by-2 contingency table shown in Table 3.6.

Using the methods outlined in Section 2 of Chapter 2, the estimated odds ratio is $(14/15)/(15/21) = 1.31$, and the standard error of its logarithm is estimated to be $\sqrt{1/14+1/15+1/15+1/21} = 0.50$, giving a 95% confidence interval (0.48, 3.56). Since this confidence interval includes 1, there could be no association between the PBC concentration and the shell thickness. This conclusion is in agreement with the result from Pearson's chi-squared test, which gives a chi-squared statistic of 0.28 and corresponding *p*-value 0.59.

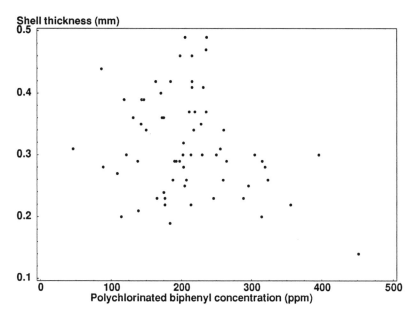

FIGURE 3.14: Scatter Plot of Data from Pelican Eggs

	PCB Concentration	
	<200 ppm	200+ ppm
<0.3 mm	14	15
Thickness 0.3+ mm	15	21

TABLE 3.6: Dichotomised Pelican Egg Data

Note that even though the association is not statistically significant, the odds ratio in the sample is greater than 1, corresponding to a positive association between the PCB concentration and the shell thickness, contrary to expectation. You would expect that a higher concentration of the pollutant would result in a lower shell thickness.

Information is lost by grouping the data. However, it is not necessary to group both the outcome and the determinant. In Section 3 of Chapter 2 you saw how to analyse data from a study in which the outcome is continuous and the determinant is dichotomous, by comparing means using the two-sample t-test. If you divide the data into two groups according to whether the PCB concentration is less or more than 200 ppm, the data may be graphed as box plots as follows.

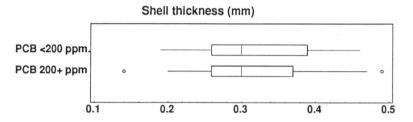

FIGURE 3.15: Shell Thickness of Pelican Eggs by Pollutant Level

There does not appear to be any difference in the distributions in the two samples, again indicating no evidence of any association between the determinant and the outcome. The sample means are 0.3166 mm in the group having the lower PCB concentration and 0.3161 mm in the group with the higher PCB concentration. The two-sample t-test gives a t-statistic of -0.02, with p-value 0.98, so the null hypothesis is not rejected.

Fitting a Straight Line

You can summarise the data in the scatter plot by fitting a straight line. In conventional statistical analysis the line fitted is the *least squares line*, which minimises the distances of the points to the line, measured in the vertical direction. This line is also called the *regression line*, and may be represented as

$$y = a + b x \tag{3.5}$$

where a is the *intercept* and b is the *slope* or *regression coefficient*.

There is a linear association between a continuous determinant and a continuous outcome if the slope is different from 0. Table 3.7 shows printout obtained from a statistical package when the pelican eggs data are analysed using linear regression.

This printout contains information about the fitted regression model. It shows that the fitted regression line is given by the equation

Thickness = 0.375 − 0.00028 PCB

where PCB is the polychlorinated biphenyl concentration. The standard error of the regression coefficient is 0.00013, giving a 95% confidence interval (−0.00054, −0.00002). The printout also gives a *p*-value for testing the null hypothesis that the population regression coefficient is 0. Since this -value is less than 0.05, the null hypothesis should be rejected: you would conclude that there is a negative association between the PCB concentration and the shell thickness.

```
Linear Regression Analysis
Response: Thickness(mm)
 Column Name          Coeff    StErr p-value        SS
```

	Column Name	Coeff	StErr	p-value	SS
0	Constant	0.37494	0.02990	0.00000	6.50329
1	PCB (ppm)	-0.00028	0.00013	0.04217	0.02649

df:63	RSq:0.06391	s:0.07848	RSS:0.38802

TABLE 3.7: Regression Analysis Printout from a Statistical Package

According to the fitted straight line model, the shell thickness decreases by 0.00028 mm for each increase of the PCB concentration by 1 ppm. But the confidence interval indicates that this rate of decrease could be as low as 0.00002 mm.

Testing Assumptions

Linear regression analysis rests on three assumptions as follows.

(1) The association is linear.

(2) The variability of the errors (in the outcome variable) is uniform.

(3) These errors are normally distributed.

These assumptions may be assessed by examining the residuals. To assess the first two assumptions, the residuals should be plotted against the *predicted values* given by the linear model. The normality assumption may be assessed by plotting the residuals against their normal scores, and tested using the Shapiro–Wilk test. This is left as an exercise.

In the scatter plot there is an outlier corresponding to the highest PCB concentration (452 ppm) and the lowest shell thickness (0.14 mm). If this outlier is omitted, the association is no longer statistically significant. The following graph shows the scatter plot of the data with the two fitted regression lines, one (the solid line) based on the data in which the outlier is retained and the other (the dotted line) when it is omitted.

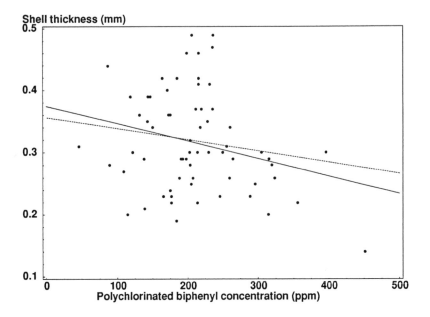

FIGURE 3.16: Data from 65 Pelican Eggs with Fitted Straight Lines

6: One-way Anova by Regression

Table 3.8 gives some data reported by Hendy and Charles (1970). These are the silver contents (expressed as percentages) of four samples of Byzantine coins discovered in Cyprus. The samples were known to have been minted at different times.

| | Sample | | |
I	II	III	IV
5.9	6.9	4.9	5.3
6.8	9.0	5.5	5.6
6.4	6.6	4.6	5.5
7.0	8.1	4.5	5.1
6.6	9.3		6.2
7.7	9.2		5.8
7.2	8.6		5.8
6.9			
6.2			

TABLE 3.8: Silver Percentages in Byzantine Coins

A question of interest here is whether the silver contents in the four mintings are different. As you saw in Section 3 of this chapter the one-way anova procedure may be used to answer this question.

Consider first the simpler question: are the first two mintings different? As you saw in Section 4 of Chapter 2 the two-sample t-test may be used. The means are 6.744 for Sample I and 8.243 for Sample II, the standard deviations are 0.543 and 1.100 respectively, and using Equations (2.5) and (2.6) the standard error of the difference between the means is 0.418. The t-statistic is 3.587 with 14 degrees of freedom, giving a p-value 0.003 for testing the null hypothesis of equal population means, so the two mintings have different silver contents.

Now consider an analysis based on fitting the simple regression model. Since the determinant (Minting I or II) is dichotomous it may be represented as a variable taking the values 0 or 1. Thus the data may be represented as (x, y) pairs, where x is the determinant (Minting I or II) and y is the outcome (the silver content), as shown below. Even though x is not a continuous variable, the methods of the preceding section apply, and the data may be displayed as a scatter plot, and a regression line may be fitted, as shown in Figure 3.17.

x	y
0	5.9
0	6.8
0	6.4
0	7.0
0	6.6
0	7.7
0	7.2
0	6.9
1	6.9
1	9.0
1	6.6
1	8.1
1	9.3
1	9.2
1	8.6

The equation of the fitted regression line is $y = 6.744 + 1.498\,x$, indicating that the outcome variable (the percent silver) increases by 1.498 when the determinant increases from 0 to 1. Since x is 0 for Minting I and x is 1 for Minting II, you can see that 1.498 is the estimate of the difference between the percentages of silver in the two mintings. This is the same as the difference between the two sample means, 6.744 and 8.243 (allowing for round-off error in the third decimal digit).

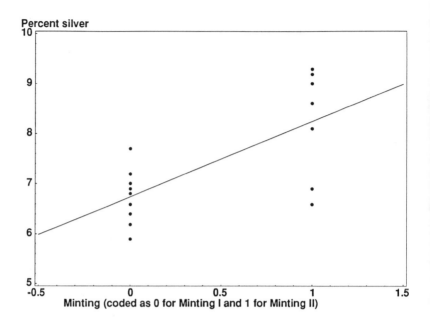

FIGURE 3.17: Percentage Silver vs Minting I or II, with Regression Lines

The following printout (Table 3.9) is obtained when a statistical package is used to do the regression analysis.

```
Linear Regression Analysis
Response: % silver
  Column Name        Coeff   StErr p-value        SS

       0 Constant    6.744   0.276  0.000   876.160
       1 Minting II   1.498   0.418  0.003     8.841

  df:14         RSq:0.479 s:0.829        RSS:9.619
```

TABLE 3.8: Regression Analysis of Silver Percentages

This printout contains the estimates of the constant and the slope parameters in the regression model, their standard errors, and a *p*-value for testing the null hypothesis of no association between the determinant and the outcome. Note that the *p*-value is the same as that obtained when the two-sample *t*-test is used. In fact the regression analysis gives precisely the same result as the two-sample *t*-test.

The bottom line of the regression analysis printout gives the number of degrees of freedom, a proportion of the squared variation in the data accounted for by the model, an estimate of the standard deviation of the errors, and the residual sum of squares.

The assumptions for the regression analysis should be checked. For a dichotomous determinant the linearity assumption is always satisfied, and the scatter plot reveals that it is reasonable to assume that the standard deviation of the errors is constant. As usual, the normality assumption may be assessed by plotting residuals against normal scores and applying the Shapiro–Wilk test.

Multicategorical Determinant

The method described above extends to data structures with a multicategorical determinant, but first it is necessary to extend the regression model (3.5) to one involving two or more determinant variables. If these variables are denoted by x_1, x_2, ..., x_c, the extended model takes the form

$$y = a + \sum_{j=1}^{c} b_j x_j \qquad (3.6)$$

The constants b_1, b_2, ..., b_c are called the *regression coefficients*. These regression coefficients measure the associations between the determinants and the outcome variable.

As an illustration of the multiple regression method, consider again the silver contents of the Byzantine coins. The data from all four samples are structured in a form suitable for regression analysis in Table 3.9. In this table the determinant variables x_1, x_2, x_3 and x_4 indicate which of the four samples the outcome variable (y) belongs to, and are known as *indicator variables*.

The following printout (Table 3.10) is then obtained when the multiple regression model is fitted to these data using a statistical package. The indicator variable corresponding to Minting I is omitted because its effect is contained in the constant term. You can see that the first two rows in the printout are similar to the results obtained from the simpler regression analysis involving just the first two samples, and the regression coefficients are the same although the standard errors and the p-value for Minting II are different.

This printout may be used to compare the populations from which the data are sampled. The comparison is made using Minting 1 as a referent population, so the regression coefficients reflect the differences between Mintings II, III and IV, respectively, and Minting I. Thus you can see that the estimated difference between the mean silver contents in Mintings III and I is −1.869% (so that Minting III has a mean 1.869% less than Minting I), and similarly the estimated difference between Mintings IV and I is −1.130%. Since the p-values for all three comparisons are small, it is reasonable to conclude that Mintings III and IV produced coins with lower silver content than

Minting I, which in turn produced coins with lower silver content than Minting II.

x_1	x_2	x_3	x_4	y
1	0	0	0	5.9
1	0	0	0	6.8
1	0	0	0	6.4
1	0	0	0	7.0
1	0	0	0	6.6
1	0	0	0	7.7
1	0	0	0	7.2
1	0	0	0	6.9
0	1	0	0	6.9
0	1	0	0	9.0
0	1	0	0	6.6
0	1	0	0	8.1
0	1	0	0	9.3
0	1	0	0	9.2
0	1	0	0	8.6
0	0	1	0	4.9
0	0	1	0	5.5
0	0	1	0	4.6
0	0	1	0	4.5
0	0	0	1	5.3
0	0	0	1	5.6
0	0	0	1	5.5
0	0	0	1	5.1
0	0	0	1	6.2
0	0	0	1	5.8
0	0	0	1	5.8

TABLE 3.9: Silver Percentages Structured for Regression Analysis

```
Linear Regression Analysis
Response: % silver
 Column Name           Coeff    StErr p-value          SS
```

Column Name	Coeff	StErr	p-value	SS
0 Constant	6.744	0.231	0.000	1162.957
2 Minting II	1.498	0.349	0.000	26.668
3 Minting III	-1.869	0.416	0.000	6.050
4 Minting IV	-1.130	0.349	0.004	5.029

```
    df:23          RSq:0.774 s:0.692          RSS:11.015
```

TABLE 3.10: Multiple Regression Analysis of Silver Percentages

It is not possible to determine from the printout whether the silver contents for Mintings III and IV are statistically different: to decide this, you would need to choose either Minting III or Minting IV as the referent population. Omitting x_4 from the regression analysis gives the following alternative printout (Table 3.11).

```
Linear Regression Analysis
Response: % silver
Column Name          Coeff    StErr  p-value            SS

       0 Constant    5.614    0.262   0.000       1162.957
       1 Minting I   1.130    0.349   0.004          0.445
       2 Minting II  2.629    0.370   0.000         35.912
       3 Minting III -0.739   0.434   0.102          1.391

   df:23          RSq:0.774 s:0.692            RSS:11.015
```

TABLE 3.11: Alternative Regression Analysis of Silver Percentages

Since the p-value corresponding to Minting III is greater than 0.05, there is insufficient evidence to conclude that the silver contents for Mintings III and IV are different. To combine these indistinguishable samples, the regression analysis may be repeated with the indicator variables omitted for both Mintings III and IV, giving the following final printout (Table 3.12).

The conclusion from this analysis is that Mintings III and IV have the same silver content, which is 1.399% lower than that of Minting I and 2.897% lower than that of Minting II. The standard errors may be used to obtain and graph confidence intervals for these differences. However, before making a definite conclusion, the statistical assumptions need to be checked, as considered next.

```
Linear Regression Analysis
Response: % silver
Column Name          Coeff    StErr  p-value            SS

       0 Constant    5.345    0.217       0       1162.957
       1 Minting I   1.399    0.323       0          0.445
       2 Minting II  2.897    0.348       0         35.912

   df:24          RSq:0.746 s:0.719            RSS:12.407
```

TABLE 3.12: Reduced Regression Model for Silver Percentages

Checking the Assumptions

In multiple regression we make the same three assumptions as in simple regression analysis, the first assumption being that the relationship between the outcome and the determinants is linear in the population. For indicator variables arising from different subpopulations this assumption is automatically satisfied (since indicator variables have only two values, which determine a straight line). The second assumption, that the standard deviation is the same in each subgroup (the *homogeneity* assumption), may be assessed by looking at box plots, as shown below in Figure 3.18.

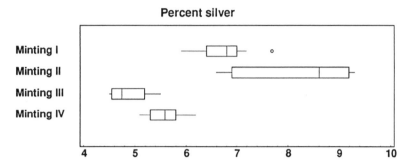

FIGURE 3.18: Box Plots of Silver Percentages

It is clear from the box plots that the spread of the data for Minting II is greater than the spreads of the other samples, so the homogeneity assumption is very dubious. If the percentages are re-expressed as logarithms, the spreads are closer, as Figure 3.19 shows. These graphs suggest that for the purposes of statistical analysis the percentages are better expressed as logarithms.

When the one-way anova procedure is applied to the log-transformed silver percentages, the F-statistic is 30.49 with 3 and 23 degrees of freedom, giving a p-value very close to 0, so the null hypothesis that the populations have a common mean should be rejected.

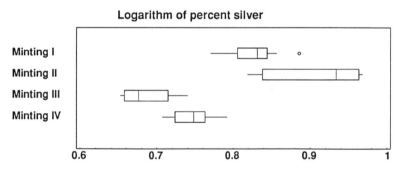

FIGURE 3.19: Box Plots of logarithms of Silver Percentages

Now consider the analysis based on fitting the multiple regression model, using Minting IV as the referent population (Table 3.13).

```
Linear Regression Analysis
Response: % silver
  Column Name          Coeff    StErr p-value            SS

         0 Constant    0.749    0.016  0.000         17.639
         1 Minting I   0.079    0.021  0.001          0.005
         2 Minting II  0.164    0.023  0.000          0.149
         3 Minting III -0.062   0.027  0.029          0.010

    df:23          RSq:0.799 s:0.042            RSS:0.041
```

TABLE 3.13: Regression Model for logarithms of Silver Percentages

This printout shows that Mintings III and IV have different silver percentages, in contrast to the result obtain with the untransformed data. So the conclusion may be expressed as

Minting III < Minting IV < Minting I < Minting II

Confidence intervals for the differences in the logarithms of the percentages may be obtained by taking two standard errors on each side of the estimated values, and these confidence intervals may be converted back to percentages by exponentiation.

The remaining assumption that needs to be checked is the normality assumption. This is left as an exercise.

Summary

Chapter 3 has continued the discussion begun in Chapter 2, looking at statistical methods for analysing data arising from comparative studies, where the objective is to investigate an association between a determinant and an outcome variable. Each variable may be measured in three different ways: dichotomous, multicategorical (with more than two categories), or continuous. This gives rise to various combinations of data types, for which different methods are needed. When the determinant is categorical the study design may be matched or unmatched, again requiring different methods for analysis. In Chapter 2 we considered the simplest situations with dichotomous determinants while Chapter 3 covered multicategorical and continuous determinants.

General Contingency Tables

You saw how the methods introduced in Chapter 2 for assessing associations in simple contingency tables may be extended to more general categorical data, and how categories having similar associations may be combined to provide simpler interpretations. These methods were illustrated with two examples, one investigating snoring as a risk factor for heart disease, and another examining the relationship between hair colour and eye colour in children.

One-way Analysis of Variance

In this section you saw how continuous outcomes in several populations may be compared by the one-way anova procedure, using survival times of patients with different cancers as an illustration. A data transformation was needed to satisfy the statistical assumptions.

Two-way Analysis of Variance

One-way anova is appropriate for independent samples. For data structured as matched sets, the samples are correlated and an adjustment is needed to remove these correlations, leading to the two-way anova procedure. You saw an application of this method to scores awarded to synchronised swimming contestants by five different judges.

Simple Regression

Data in which both the outcome and the determinant are continuous variables may be displayed as a scatter plot and summarised by fitting a straight line. The slope of this fitted line is a measure of the association between the two variables. This method, called simple linear regression, was illustrated using data from a study investigating the association between shell thickness and the level of an industrial pollutant in Anacapa pelican eggs, the results being sensitive to an outlier in the data.

One-way Anova by Regression

Chapter 3 concluded by showing how the simple regression model may be extended to handle a multicategorical determinant, providing a more informative alternative to one-way anova for comparing continuous outcomes from several populations. Silver contents of samples of Byzantine coins from Cyprus were used to illustrate the method.

Exercises

Exercise 3.1: Neyzi et al (1975) reported the following data from a study of the association between parental socio-economic status and breast development in 12-13 year old Turkish girls. Analyse these data using the methods outlined in Section 2.

		Breast development				
		none	tiny	small	moderate	full
	1	2	14	28	40	18
Parent	2	1	21	25	25	4
SEC	3	1	12	12	12	2
	4	6	17	34	33	6

Exercise 3.2: Use the one-way anova procedure to analysis the weight gains for all three groups of anorexic girls. (See Exercise 2.5.)

Exercise 3.3: The following data were obtained from a study by Mazess et al (1984) investigating a new method of measuring body composition, and comprise the body fat percentage, age and sex for 18 normal adults aged between 23 and 61 years. Write a statistical report on these data, including

(a) a comparison of the ages and the fat percentages of the males and the females, using 95% confidence intervals and *t*-tests;

(b) a scatter plot of the fat percentages versus age;

(c) a regression analysis modelling the association between fat percentage and age;

(d) an assessment of the statistical assumptions underlying the regression analysis.

Subject	age	%fat	sex
1	23	9.5	M
2	23	27.9	F
3	27	7.8	M
4	27	17.8	M
5	39	31.4	F
6	41	25.9	F
7	45	27.4	M
8	49	25.2	F
9	50	31.1	F
10	53	34.7	F
11	53	42.0	F
12	54	29.1	F
13	56	32.5	F
14	57	30.3	F
15	58	33.0	F
16	58	33.8	F
17	60	41.1	F
18	61	34.5	F

Exercise 3.4: Berry (1987) analysed resistances of five different types of electrodes applied to the arms of 16 subjects to see whether all electrode types performed similarly. Analyse these data (listed below) using two-way anova.

| | | Electrode type | | | |
Subject	1	2	3	4	5
1	500	400	98	200	250
2	660	600	600	75	310
3	250	370	220	250	220
4	72	140	240	33	54
5	135	300	450	430	70
6	27	84	135	190	180
7	100	50	82	73	78
8	105	180	32	58	32
9	90	180	220	34	64
10	200	290	320	280	135
11	15	45	75	88	80
12	160	200	300	300	220
13	250	400	50	50	92
14	170	310	230	20	150
15	66	1000	1050	280	220
16	107	48	26	45	51

Exercise 3.5: Use the multiple regression model to compare the survival times of the cancer patients given in Section 3. Combine any groups that are not statistically different. Give confidence intervals for the mean survival times (in days) in each (combined) group.

References

Berry, D.A. (1987): Logarithmic transformations in ANOVA, *Biometrics*, **43**, pages 439–456.

Cameron, E. & L. Pauling (1978): Supplemental ascorbate in the supportive treatment of cancer: re-evaluation of prolongation of survival times in terminal human cancer, *Proceedings of the National Academy of Science USA*, **75**, pages 4538–4542.

Fligner, M.A. & J.S. Verducci (1988): A nonparametric test for judges' bias in an athletic competition, *Applied Statistics*, **37**, pages 101–110.

Goodman L.A. (1981): Association models and canonical correlation in the analysis of cross-classifications having ordered categories, *Journal of the American Statistical Association* **76**, pages 320–334.

Hendy, M.F. & J.A. Charles (1970): The production techniques, silver content and circulation history of the twelfth-century Byzantine Trachy, *Archaeometry*, **12**, pages 13–21.

Mazess, R.B., W.W. Peppler & M. Gibbons (1984): Total body composition by dual-photon (^{153}Gd) absorptiometry, *American Journal of Clinical Nutrition*, **40**, pages 834–839.

Neyzi, O., H. Alp & A. Orhon (1975): Breast development of 318 12–13 year old Turkish girls by socio-economic class of parents, *Annals of Human Biology*, 2, **1**, pages 49–59.

Norton, P.G. & E.V. Dunn (1985): Snoring as a risk factor for disease: an epidemiological survey, *British Medical Journal*, **291**, pages 630–632.

Risebrough, R.W. (1972): Effects of environmental pollutants upon animals other than man, *Proceedings of the 6th Berkeley Symposium on Mathematics and Statistics*, pages 443–463. University of California Press.

Winer, B.J. (1971): *Statistical Principles in Experimental Design* (2nd ed.). McGraw-Hill. New York.

4

MANTEL–HAENSZEL METHODS

1: Introduction

Chapters 2 and 3 cover basic statistical methods for epidemiological research, emphasising situations in which the outcome is measured on a continuous scale. Epidemiology is more often concerned with dichotomous outcomes, such as the presence or absence, rather than the degree of severity, of a health condition. In this chapter we focus on methods for handling such binary data. The primary focus is on the odds ratio, which is a valid measure of association in a cross-sectional, cohort, or case-control study.

Bias due to confounding can arise due to the presence of additional determinants which are related to the determinant of interest. The present chapter is largely concerned with a particular set of methods, first suggested by Mantel and Haenszel (1959) and consequently named after these authors, for dealing with confounding in two-by-two tables. These methods essentially involve analysing a study by subdividing the data into sets of similar components and then combining the component sets by a kind of averaging process. While these traditional methods have been superseded to some extent by the logistic and Poisson regression models covered in Chapters 5 and 6, they remain important because they can provide the investigator with insight into the patterns of variation in the data.

To set the scene consider the very simplest situation, in which both the outcome and the determinant are dichotomous and there are no covariates, so that the data are structured as a two-by-two table of counts, as follows.

		Exposure	
		E	$\bar{E}$
Outcome	D	a	b
	$\bar{D}$	c	d

The outcome (often a 'disease') and the determinant (often an 'exposure') are designated by the symbols D and E, respectively, so a and b are the numbers of 'exposed' and 'nonexposed' subjects who experience the outcome, and c and d are the numbers who do not, respectively. Following the classification of study types given in Chapter 1, the data may arise from a cross-sectional study, a cohort study, or a case-control study.

Measures of Association

If the data arise from a cross-sectional study or from a cohort study, then the exposed subjects constitute a sample from the exposed individuals in the target population, so an unbiased estimate of the risk or probability of the outcome given the exposure is $R(E) = a/(a+c)$. Similarly, an estimate of this risk in the nonexposed population is $b/(b+d)$.

These estimates of risk are not valid if the data have been obtained from a case-control study, for in that case the criteria for selecting the cases (those subjects experiencing the outcome) and the controls (subjects not experiencing the outcome) are not necessarily the same. For example, if the study included all the subjects in a population having the outcome, but only 10% of the subjects in the same population who are exposed but who do not experience the outcome, the estimated risk of the outcome given the exposure is given by $a/(a+10c)$ rather than $a/(a+c)$.

The risks may be compared in alternative ways. An estimate of the risk difference, or *excess risk*, associated with the exposure is $a/(a+c) - b/(b+d)$. This is just a difference in proportions, for which some basic methods for statistical analysis have been given in Chapter 2. The second measure of comparison is the risk ratio or *relative risk* (*RR*) in the exposed and unexposed groups. An estimate of *RR* based on the simple contingency table is $(a/(a+c))/b/(b+d))$. Methods for the analysis of relative risks are presented in Section 4 of this chapter.

The size of the relative risk is useful for classifying risk factors. However, if the risk in the nonexposed population is close to zero, a relative risk may be large, but still not important. Suppose for the sake of argument that the risk of dying in a plane crash travelling on a particular flight on Airline A is one in ten million, compared with one in a million on Airline B, so $RR = 10$. Given the choice, which airline would you choose to fly? Airline A of course! But what if Airline B has substantially better service and cheaper tickets than Airline A? This sizable increase in benefit needs to be weighed against the miniscule increase in absolute risk, and in this case it is the excess risk that is important, rather than the relative risk.

The odds ratio is similar to the relative risk, but is defined in terms of odds rather than probabilities, and has advantages over the other measures. Unlike the relative risk, which cannot be measured from a case-control study, the odds ratio is a valid measure of risk association in a case-control, cohort, or cross-sectional study.

As you saw in Chapter 2, the odds of an outcome occurring in the exposed group E is defined as the risk ratio $O(E) = R(E)/(1-R(E))$. Assuming that the data in the contingency table having the counts a, b, c and d give valid estimates of the risks, the estimated odds in the exposed group is thus a/c. Similarly, an estimate of the odds in the unexposed group is b/d, and the estimated odds ratio is thus

$$OR = \frac{ad}{bc} \tag{4.1}$$

Equation (2.3) gives an asymptotic formula for the standard error of the logarithm to base e (or *natural* logarithm, denoted by ln) of an observed odds ratio, namely

$$SE\,[\ln(OR)] = \sqrt{\frac{1}{a} + \frac{1}{b} + \frac{1}{c} + \frac{1}{d}} \tag{4.2}$$

If any of these counts is 0, the above formulas do not give very useful results, and in such cases it is customary to add a small amount to the zero count (such as 0.5 or 0.25) to ensure that the estimates are finite.

In Chapter 2 Pearson's chi-squared statistic was suggested for testing the null hypothesis that the population odds ratio in a two-by-two table is 1. Epidemiologists often use another chi-squared test for this null hyothesis, called the *Mantel–Haenszel test*. The formula for this test statistic is

$$\chi^2_{MH} = \frac{(n-1)(ad-bc)^2}{(a+c)(b+d)(a+b)(c+d)} \tag{4.3}$$

where $n = a+b+c+d$, and the p-value is again obtained from the chi-squared distribution with 1 degree of freedom. Pearson's test is very similar to the Mantel–Haenszel test: the only difference is in the numerator, where Pearson's test has n and the Mantel–Haenszel test has $n-1$. The tests are thus asymptotically equivalent.

Advantage of Odds Ratio

Consider some data cited by Kleinbaum et al (1982, pages 9–12), taken from a cohort study of 609 white males aged between 40 and 76, free of coronary heart disease (CHD) and living in Evans County, Georgia in 1960, and followed up for seven years. During this period 71 new cases of CHD were observed in the cohort. An exposure variable of interest was serum cholesterol level (CHL), a subject being considered 'exposed' if his initial level was above 260 mg/100 mL. These data are tabulated as follows:

$$CHL > 260 \ mg/100 \ mL$$

		Yes	No
	Yes	14	57
CHD	No	91	447
		105	504

Using Equations (4.1) and (4.2), the estimated log odds ratio is the natural logarithm of $(14 \times 447)/(57 \times 91)$, or 0.187, with standard error 0.32, giving a 95% confidence interval (0.64, 2.3) for the odds ratio (using the two standard error rule). Equation (4.3) gives a chi-squared statistic of 0.345 and p-value 0.56 for the Mantel–Haenszel test that the population odds ratio is 1. The conclusion is that there is insufficient evidence of an association, and the population odds ratio could be as low as 0.64 and as high as 2.3.

Recall that Equation (4.1) is obtained as the ratio of the components a/c and b/d, which are valid estimates of the odds in the exposed and nonexposed groups in a cohort or cross-sectional study, but not in a case-control study. However, Equation (4.1) also provides a valid estimate of the odds *ratio* in a case-control study. To see why this is so, consider a case-control study in which π_1 and π_0 are the proportions of cases and controls, respectively, selected from the target population for inclusion in the study. Given observed numbers of exposed and nonexposed cases a and b, respectively, in the study, the corresponding estimated numbers in the target population are a/π_1 and b/π_1. Similarly, the estimated numbers of exposed and nonexposed controls in the target population are c/π_0 and d/π_0, respectively. Using these estimates to calculate probabilities, the odds of outcome are thus $(a/\pi_1)/(c/\pi_0)$ for an exposed subject, and $(b/\pi_1)/(d/\pi_0)$ for a nonexposed subject. The ratio of these odds is $(ac)/(bd)$, in agreement with Equation (4.1).

The Evans County cohort study may be used to illustrate this concept. Suppose you wish to investigate the CHL/CHD association by undertaking a case-control study at the end of the seven years of follow-up, by which time 71 cases of CHD have occurred. Accordingly you select 71 representative subjects from the 538 who had not suffered the disease and determine their serum cholesterol level at entry to the study, giving the table shown on the right in the next exhibit:

Cholesterol level > 260 mg/100 mL

		Yes	No		Yes	No
	Yes	14	57	Case	14	57
CHD						
	No	91	447	Control	12	59

Cohort Study   Case-Control Study

Based on the cohort study in which all subjects are selected, the risk of CHD in the exposed group is $14/(14+91) = 0.133$, compared with $57/(57+447)$, or 0.113 in the nonexposed group. When these calculations are applied to the data from the case-control study, the totally erroneous estimates of $14/(14+12) = 0.56$ and $57/(57+59) = 0.49$ are obtained for the risks. However, the case-control study gives a valid estimate of the odds ratio, namely $(14 \times 59)/(57 \times 12) = 1.21$.

The odds ratio is a valid measure of association between a determinant and an outcome, irrespective of study type, and the methods for analysing odds ratios apply to case-control studies as well as cross-sectional studies and cohort studies. In particular, Equation (4.2) may be used to get a confidence interval for an odds ratio based on a case-control study. Using the data from the above example, the case-control study gives a 95% confidence interval of (0.51, 2.88) for the odds ratio. This compares with the narrower interval (0.64, 2.3) based on the cohort study.

2: Confounding in 2-by-2 Tables

The data in Table 4.1 are assumed to have come from a cross-sectional study involving 715 consecutive births in a hospital, cited by Bishop (1969). The outcome is the baby failing to survive for one month after birth, and the exposure is the amount of prenatal care (PNC) received by the mother (less than one month, or at least one month). These data are also classified by a third binary variable, the prenatal care clinic (A or B).

Prenatal care < 30 days

		Yes	No		Yes	No
	Yes	3	4	Yes	17	2
Died within 1 month						
	No	176	293	No	197	23

Clinic A   Clinic B

TABLE 4.1: Neonatal Mortality vs Prenatal Care in Two Clinics

The estimated odds ratios are $(3 \times 293)/(4 \times 176) = 1.25$ (95% CI: 0.27 – 5.83) for Clinic A and $(17 \times 23)/(2 \times 197) = 0.99$ (95% CI: 0.21 – 4.71) for Clinic B. Neither of these estimates is significantly different from 1, so there is no evidence of an association between prenatal care and neonatal mortality in either of the two clinics.

This conclusion differs markedly from that obtained by aggregating the data from the two clinics, where the estimated odds ratio is $(20 \times 316)/(6 \times 373) = 2.824$ with 95% confidence interval (1.10, 7.25).

		Exposure	
		Prenatal Care (days)	
		0-29	30+
	Died	20	6
Outcome			
	Survived	373	316

The conclusion is that neonatal death is more likely in women attending Clinic B, and the amount of prenatal care is not an issue. The clinic is a confounder in the association between prenatal care and neonatal mortality. As you can see from this example, failure to take account of a confounder can give a misleading result.

We now give four artificial but instructive examples which illustrate how confounding arises, and we then show how *stratification* may be used to control for it.

Confounding Examples

Suppose you have two 2-by-2 tables relating an exposure E to an outcome D, one for each level of another dichotomous variable F (a covariate). The (hypothetical) data given in the following tables (Table 4.2) provide a concrete example (Example 1).

F			$\overline{F}$			Aggregated		
	E	$\overline{E}$		E	$\overline{E}$		E	$\overline{E}$
D	50	30	D	30	20	D	80	50
$\overline{D}$	40	60	$\overline{D}$	90	130	$\overline{D}$	130	190
OR = 2.5			OR = 2.17			OR = 2.34		

TABLE 4.2: Confounding Example 1 (No Confounding)

The odds ratios in the strata are slightly different (2.5 and 2.17), but both are statistically different from 1 (the p-values for the

Mantel–Haenszel test of association are 0.003 and 0.014, respectively) and when the data are aggregated, the estimated odds ratio (2.34) is close to the average of the separate odds ratios. The p-value given by the Mantel–Haenszel test based on the pooled data is less than 0.001.

The result is not unexpected. The strata could refer to two different studies investigating the same question. Each study gives a statistically significant association, and pooling the data from the two studies gives a more accurate estimate of the population odds ratio.

In Example 2, shown below in Table 4.3, there is no association in each stratum, but a significant association appears when the data are aggregated. The odds ratio in each stratum is 1, indicating no association between E and D, but the overall odds ratio in the combined table is 2.1 (95% confidence interval 1.36, 3.23), indicating a positive association. This example is similar to the study of prenatal care and neonatal mortality in the two clinics: if the covariate is ignored there appears to be an association between the exposure and the outcome, but the association disappears when the data are broken down by the levels of the covariate.

	F			$\overline{F}$			Aggregated	
	E	$\overline{E}$		E	$\overline{E}$		E	$\overline{E}$
D	80	20	D	10	30	D	90	50
$\overline{D}$	80	20	$\overline{D}$	40	120	$\overline{D}$	120	140
	OR = 1			OR = 1			OR = 2.1	

TABLE 4.3: Confounding Example 2 (Inflation)

Stratifying by a covariate can also unmask an association, as shown in the next two examples. In Example 3 (Table 4.4) the odds ratios are similar (both close to 2) and significantly greater than 1 in each subgroup (the Mantel–Haenszel test for no association gives p-values 0.032 and 0.008, respectively) but the association disappears completely when the data are pooled.

	F			$\overline{F}$			Aggregated	
	E	$\overline{E}$		E	$\overline{E}$		E	$\overline{E}$
D	120	10	D	80	190	D	200	200
$\overline{D}$	160	30	$\overline{D}$	40	170	$\overline{D}$	200	200
	OR = 2.25			OR = 1.79			OR = 1	

TABLE 4.4: Confounding Example 3 (Masking)

The fourth example (Table 4.5) shows a similar effect, but in this case the odds ratios in the two subtables are markedly different.

	F			$\overline{F}$			Aggregated	
	E	$\overline{E}$		E	$\overline{E}$		E	$\overline{E}$
D	155	20	D	45	180	D	200	200
$\overline{D}$	180	80	$\overline{D}$	20	120	$\overline{D}$	200	200
	OR = 3.44			OR = 1.5			OR = 1	

TABLE 4.5: Confounding Example 4 (Effect Modification)

Figure 4.1 graphs the 95% confidence intervals for the odds ratios in the separate and combined strata for each of the four examples. In Example 1 the odds ratio is not distorted by the covariate, so there is no confounding. In Examples 2 and 3 the covariate F distorts (either by inflation or masking) the association between the exposure and the outcome, and consequently each is a straightforward example of confounding. Example 4 is more complicated, because the association is different in the two strata. This situation is called *effect modification* rather than simple confounding.

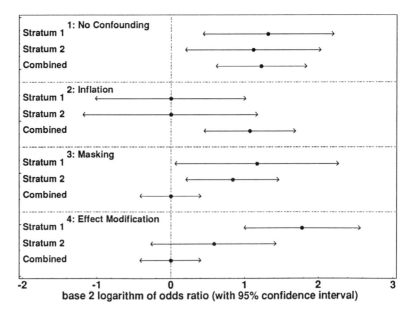

FIGURE 4.1: Four Examples Illustrating Confounding

Confounding can arise when there is an association between two risk factors. For Example 1 (from Table 4.2), the association between E and F for each outcome category is shown in the following table. Neither odds ratio is significantly different from 1, so the two risk factors are not associated.

		D			$\bar{D}$	
		E	$\bar{E}$		E	$\bar{E}$
Example 1	F	50	30	F	40	60
	$\bar{F}$	30	20	$\bar{F}$	90	130
		OR = 1.11			OR = 0.96	

Now look at the same analysis for Example 2 (from Table 4.3). In this case the odds ratio for the association between E and F in each outcome category is 12, which is very high. The odds ratios for the associations between E and F in Examples 3 and 4 are even higher.

		D			$\bar{D}$	
		E	$\bar{E}$		E	$\bar{E}$
Example 2	F	80	20	F	80	20
	$\bar{F}$	10	30	$\bar{F}$	40	120
		OR = 12			OR = 12	

These examples demonstrate that confounding can arise when two or more risk factors are mutually associated. For this reason it is important to give measures of the associations between the various risk factors of interest when reporting the results from an epidemiologic study.

Correlation between two determinants is a necessary condition for confounding, but it is not a sufficient condition. A covariate must also be an independent risk factor for the outcome before it can be a confounder.

Ignoring a confounder gives a single *crude* estimate of an odds ratio. To obtain a single estimate of an odds ratio *adjusted* for a confounding variable, the individual estimates obtained from each strata need to be combined in some way. It is also important to check that individual estimates are compatible before combining them. Methods for testing the homogeneity of odds ratios and then combining them are discussed in the next section.

3: Combining Odds Ratios

If you have several two-by-two tables corresponding to different strata, a measure of the overall association may be obtained by combining the statistics obtained from the component tables. Since odds ratios have skewed distributions, it is better to work with their logarithms. An estimate for the overall log odds ratio could be obtained simply by taking the average of the individual components. Take Example 3 from the preceding section. In this case the logarithms of the odds ratios in the two strata are 0.811 and 0.582 and their average is 0.696, and exponentiating this value gives 2.01 for the overall odds ratio.

Averaging the logarithms of the component odds ratios gives equal weight to each component, which is reasonable when the strata are of similar size. However, if the strata contain different numbers of subjects it is better to used a *weighted* average, with greater weight being attached to strata with larger numbers of subjects. Denoting by y_g the estimated log odds ratio in stratum g, a formula for the weighted average is

$$\bar{y} = \frac{\sum w_g y_g}{\sum w_g}, \quad w_g = \frac{1}{SE_g^2} \tag{4.4}$$

where estimates for the standard errors are given by Equation (4.2). The weights w_g, being inversely proportional to the squares of these standard errors, are thus directly proportional to the counts. There is also a formula for the standard error of the weighted average, namely

$$\bar{SE} = \frac{1}{\sqrt{\sum (1/SE_g)^2}} \tag{4.5}$$

The above statistical formulas are quite general and may be used in any situation where you need to combine estimates of a common parameter obtained from independent sources. For the data in Example 3, Equation (4.2) gives standard errors of 0.385 and 0.221 for the strata, and the results based on Equation (4.4) are as follows:

g	y_g	SE_g	w_g	$w_g y_g$
1	0.811	0.385	6.76	5.48
2	0.582	0.221	20.56	11.96
			27.32	17.44

The resulting estimate of the weighted mean is 17.44/27.32 = 0.638 and, using Equation (4.5), its standard error is $1/\sqrt{27.32} = 0.191$. The estimate of the common odds ratio is thus exp(0.638) = 1.89, with 95% confidence interval exp(0.638 $\pm$ 2 $\times$ 0.191), that is, (1.29, 2.77).

A problem with using Equation (4.4) for combining odds ratios is that if the strata are too fine, some of the counts may be very small, in which case the component odds ratios and standard errors vary considerably, giving an inaccurate result. To overcome this problem Mantel and Haenszel (1959) suggested using the estimate

$$OR_{MH} = \frac{\Sigma a_g d_g / n_g}{\Sigma b_g c_g / n_g} \qquad (4.6)$$

where g specifies the stratum. Accordingly this estimate is known as the *Mantel–Haenszel estimate* of the adjusted odds ratio. A confidence interval for the adjusted odds ratio may be obtained by using a formula given by Robins et al (1986), that is

$$SE[\ln(OR_{MH})] = \sqrt{\frac{\Sigma P_g R_g}{2R_+^2} + \frac{\Sigma(P_g S_g + Q_g R_g)}{2R_+ S_+} + \frac{\Sigma Q_g S_g}{2S_+^2}} \qquad (4.7)$$

where

$$R_+ = \Sigma R_g, \ S_+ = \Sigma S_g$$

and

$$P_g = \frac{a_g + d_g}{n_g}, \ Q_g = \frac{b_g + c_g}{n_g}, \ R_g = \frac{a_g d_g}{n_g}, \ S_g = \frac{b_g c_g}{n_g}$$

For the study of prenatal care and neonatal mortality stratified by clinic Equation (4.6) gives the Mantel–Haenszel estimate

$$\frac{3 \times 293/476 + 17 \times 23/239}{4 \times 176/476 + 2 \times 197/239} = 1.114$$

The standard error estimate calculated from Equation (4.7) is

$$\sqrt{\frac{1.422}{2 \times 3.483^2} + \frac{1.195 + 2.061}{2 \times 3.483 \times 3.128} + \frac{1.933}{2 \times 3.128^2}} = 0.554$$

so a 95% confidence interval for the adjusted odds ratio is 1.114 times exp($\pm$ 2 $\times$ 0.554), that is, (0.37, 3.37). This confidence interval may be contrasted with the interval (1.10, 7.25) calculated for the crude odds ratio which takes no account of the clinic. In light of the additional information, the conclusion is that while there is no evidence that less prenatal care is a risk factor for infant mortality, the study is inconclusive because the true odds ratio is likely to be anywhere between 0.37 and 3.37.

Applying Equation (4.6) to the data from Example 3 gives an estimated common odds ratio of 1.90 which is close to the value 1.89 given by Equation (4.4). The standard error given by Equation (4.7) is 0.191, agreeing to three decimal places with the value given by Equation (4.5).

Mantel and Haenszel (1959) also gave a statistic for testing the null hypothesis that the overall odds ratio in several two-by-two tables is 1. This test statistic should be compared with the chi-squared distribution with 1 degree of freedom, and is given by

$$\chi^2_{MHS} = \left(\sum_{g=1}^{G} \frac{a_g d_g - b_g c_g}{n_g} \right)^2 / \sum_{g=1}^{G} \frac{(a_g + c_g)(b_g + d_g)(a_g + b_g)(c_g + d_g)}{(n_g - 1)n_g^2} \qquad (4.8)$$

This generalises Equation (4.3). Applying Equation (4.8) to the data in Example 3 gives the value 11.46 for the chi-squared statistic, and the corresponding p-value 0.0007.

Can odds ratios from different strata always be combined? Only if they are all estimating the same population parameter. It seems reasonable to combine the odds ratios in Example 3 (2.25 and 1.79) but not in Example 4 (3.44 and 1.5), In general how can you decide that two or more odds ratios are compatible?

Provided the counts in the strata are sufficiently large to use the normal approximation, a chi-squared homogeneity test may be used, and is computed as follows. First, adjust the estimates of the log odds ratios to have zero weighted average (by subtracting the weighted average from each estimate). Then scale the adjusted estimates by dividing each by its estimated standard error, and calculate the sum of the squares of these z-scores. A p-value is obtained by comparing the result with a chi-squared distribution with $G - 1$ degrees of freedom, where G is the number of strata.

If some counts are small more stable estimates are needed for the components of the z-scores. More robust estimates of the standard errors are obtained by replacing the observed counts by expected counts based on fitting the overall Mantel–Haenszel odds ratio to each subtable.

Consider the application of these methods to Example 3. In this case the component z-scores are $z_1 = (0.811 - 0.638)/0.385 = 0.449$ and $z_2 = (0.582 - 0.638)/0.221 = -0.253$, giving the sum of squares 0.266. With two strata, the number of degrees of freedom for the chi-squared test is 1, so the p-value is 0.61 in agreement with the homogeneity assumption. Turning to Example 4, Equations (4.1) and

(4.2) give log odds ratios 1.237 and 0.405 with standard errors 0.273 and 0.293, respectively, so the calculated z-scores are computed to be $(1.237 - 0.851)/0.273 = 1.414$ and $(0.405 - 0.851)/0.293 = -1.522$, giving the chi-squared statistic 4.316. The corresponding p-value is 0.038, indicating lack of homogeneity.

To summarise these methods, the stratified Mantel–Haenszel estimate given by Equation (4.6) is used to combine odds ratios from different strata, and a standard error formula is available from which a confidence interval may be computed. Provided none of the counts is too small, you should first test that the components come from the same population by forming a sum of squares of z-scores and comparing this with a chi-squared distribution with one less degree of freedom than the number of strata. If you cannot reject this null hypothesis, you may then test that the overall odds ratio is 1 by using the stratified Mantel–Haenszel test.

Consider an application of these procedures to some data from a cross-sectional study cited by Freedman et al (1978, page 14). The numbers of men and women admitted in 1973 to the graduate programme in the six largest departments at the University of California, Berkeley, were as shown in the following table (Table 4.6).

	Men	Women
Accepted	1198	557
Rejected	1493	1278

TABLE 4.6: Berkeley Graduate Admissions in 1973

The estimated odds of admission for a man is $a/c = 1198/1493$, or 0.80, compared with $b/d = 557/1278 = 0.44$ for a woman. Using Equation (4.1) the estimated odds ratio is 1.84, and Equation (4.2) gives a 95% confidence interval (1.62, 2.09), which appears to indicate discrimination against women.

To find out which departments were most guilty of sexual discrimination (or to see if there is a bias due to confounding), Table 4.7 shows the same data broken down by department. The estimated odds ratios are given at the right of the table.

Note that all but two of these odds ratios are less than 1, so the bias is *in favour of women*! In fact the 95% confidence interval for the odds ratio in Department A is (0.21, 0.58), while those for the other departments straddle 1, indicating no discrimination either way. The explanation is that four of the departments (C, D, E and F) had much lower admission rates than the other two, and the women tended to apply for admission to these departments rather than the ones with the higher admission rates (A and B).

		Accepted		Rejected		
		Men	Women	Men	Women	OR
	A	512	89	313	19	0.35
	B	353	17	207	8	0.80
	C	120	202	205	391	1.13
Dept.	D	138	131	279	244	0.92
	E	53	94	138	299	1.22
	F	22	24	351	317	0.83

TABLE 4.7: Berkeley Graduate Admissions Stratified by Department

Now let us apply the methods for testing homogeneity and for combining odds ratios to these data. The relevant statistics are listed in Table 4.8, where Equation (4.4) is the basis for the computations. Substituting the relevant values from the table into Equation (4.4) gives a weighted average −0.075, and the z-scores are then computed by subtracting this average from each log odds ratio and dividing the result by the corresponding standard error.

Department	ln(OR)	SE	w	z
A	−1.052	0.263	14.49	−3.72
B	−0.220	0.438	5.22	−0.33
C	0.125	0.144	48.26	1.39
D	−0.082	0.150	44.32	−0.05
E	0.200	0.200	24.94	1.37
F	−0.189	0.305	10.74	−0.37

TABLE 4.8: Inhomogeneity of Odds Ratios for Berkeley Admissions Data

The sum of squares of the z-scores is 17.90, and comparing this with a chi-squared distribution with 5 degrees of freedom, the p-value is 0.003, so it is unreasonable to combine the separate estimates.

The log odds ratios are graphed with 95% confidence intervals

in Figure 4.2 (changing to base 2 logarithms to facilitate interpretation). Clearly Department A is incompatible with the others.

Department

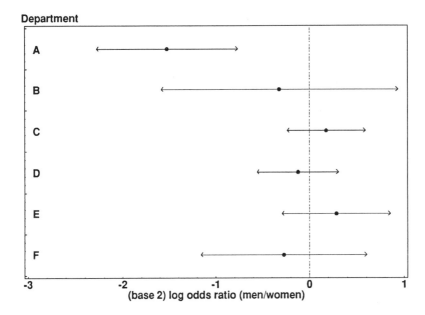

FIGURE 4.2: Odds ratios by Department: Berkeley Graduate Admissions

Excluding Department A, the process may be repeated with the
other five departments, giving a weighted average log odds ratio of
0.032, a sum of squared scores of 2.55, and a p-value of 0.635, so it is
reasonable to conclude that the odds ratios for the other five departments are the same. Since none of the counts is small, the asymptotic
formulas may be used for combining the odds ratios. From Equation
(4.5), the standard error of the weighted average is 0.087. A 95% confidence interval for the combined estimate of the log odds ratio is thus
$0.032 \pm 2 \times 0.087$, that is, $(-0.142, 0.206)$. Exponentiating these limits
gives the 95% confidence interval $(0.87, 1.23)$ for the odds ratio. The
conclusion is that the overall odds ratio for departments B, C, D, E
and F is not different from 1.

The odds ratio is an important measure of association in epidemiology because it is measurable from a case-control study, even
though the risk or prevalence of the outcome event is not directly
measurable. Cross-sectional studies are more informative than case-
control studies in the sense that they allow direct measurement of the
risk of the event, from which the relative risk corresponding to two
exposure groups may be estimated. Cohort studies are more informative still: since they involve follow-up over time they provide infor-

mation about the incidence of an outcome of interest.

The next section is concerned with methods for estimating and comparing the risks of outcomes in cross-sectional studies and cohort studies, and we focus on the relative risk as the parameter of interest.

4: The Relative Risk

For a cross-sectional or cohort study in which both the exposure and outcome are dichotomous and there are no covariates, the data may be presented as a two-by-two table of counts as follows:

		Exposure	
		E	$\overline{E}$
Outcome	D	a	b
	$\overline{D}$	c	d
		n_1	n_0

The marginal totals n_1 and n_0 refer to the total numbers of subjects in the exposed and nonexposed groups, respectively. Estimates of the risks of the outcome in the two groups are a/n_1 in the exposed group and b/n_0 in the nonexposed group. The estimated relative risk is thus

$$RR = \frac{a/n_1}{b/n_0} \qquad (4.9)$$

For a cohort study, a and b are the numbers of incident cases in the two exposure groups, and the term *cumulative incidence ratio* is used to describe this relative risk. A confidence interval for RR is obtainable using a similar formula to that for the odds ratio. If the sample size is reasonably large, the estimated standard error of the natural logarithm of relative risk is

$$SE\,[\,\ln(RR)\,] = \sqrt{\frac{1}{a} - \frac{1}{n_1} + \frac{1}{b} - \frac{1}{n_0}} \qquad (4.10)$$

A p-value for testing the null hypothesis that the relative risk is 1 is found by using the same approach as for testing an odds ratio, that is, using either the z-statistic obtained by dividing the estimate of the logarithm of the relative risk by its standard error, or the (asymptotically equivalent) Mantel–Haenszel test given by Equation (4.3).

If the risks are small the relative risk is close to the odds ratio, so Equations (4.9) and (4.10) give the same results as Equations (4.1) and (4.2) in the limit as a/n_1 and b/n_0 tend to zero.

To demonstrate the use of Equations (4.9) and (4.10), consider again two of the examples for which odds ratios were estimated in Section 2. For the association between (less) prenatal care and neonatal mortality, the estimate of the relative risk from Equation (4.9) is $(20/393)/(6/322) = 2.73$, and Equation (4.10) gives 0.459 for the estimated standard error. A 95% confidence interval for the RR is thus $2.73\exp(\pm 2 \times 0.459)$, that is, (1.09, 6.84). This may be compared with the estimated odds ratio 2.82 with 95% confidence interval (1.10, 7.25) obtained earlier. For the cohort study investigating a link between cholesterol and CHD, the estimated cumulative incidence ratio is 1.18 and the standard error estimate is 0.278, giving a 95% confidence interval (0.68, 2.06). Again this may be compared with the estimate 1.21 and 95% confidence interval (0.64, 2.3) obtained earlier for the odds ratio.

In general, the estimate for the relative risk is closer to 1 than the estimated odds ratio based on the same data, and the confidence interval for the relative risk is shorter than the equivalent confidence interval for the odds ratio based on the same data.

The Incidence Density Ratio

In practice cohort studies are complicated by the fact that the duration of exposure may differ from subject to subject. Subjects may depart before the end of the follow-up period, or may enter the study at different times. The exposed group may have been followed up for a different length of time to the control (nonexposed) group. For this reason it may be difficult to measure the total numbers (n_1 and n_0) at risk in the exposure groups. A more general parameter of interest in a cohort study is the *incidence density*, defined for each exposure group as the number of incidences of the outcome event (a for the exposed group and b for the control group) divided by the total duration of follow-up for all subjects in the group. This total duration of follow-up is thus expressed as person-times (e.g., person-years or person-days). The data layout for a cohort study is then as shown in the following table:

	Exposure	
	E	$\overline{E}$
New cases	a	b
Total Follow-up	L_1	L_0

The relative risk in the two exposure groups may now be defined as the ratio of the incidence densities in the exposed and non-exposed groups: this is called the *incidence density ratio (IDR)*. An estimate of *IDR* is thus

$$IDR = \frac{a/L_1}{b/L_0} \tag{4.11}$$

Provided the individual risks are small, an estimate of the standard error of the logarithm of the estimated incidence density ratio is given by the formula

$$SE[\ln(IDR)] = \sqrt{\frac{1}{a} + \frac{1}{b}} \tag{4.12}$$

and this may be used to obtain both a confidence interval for the incidence density ratio and a *p*-value for testing that this relative risk is 1.

Equation (4.12) also arises as the limiting case of Equation (4.10) when the risk estimates a/n_1 and b/n_0 tend to zero. The values of the person-times of follow-up L_1 and L_0 do not enter into the formula for the standard error of the relative risk estimate, and only their ratio enters into the formula for the relative risk. It follows that only the expected incidence rates in the two groups are needed. These could be obtained from an independent source.

As a hypothetical illustration, suppose that 50 motor accident fatalities are recorded in a city in the 12 month period immediately after the introduction of random blood alcohol testing of motorists, compared with 75 fatalities in the previous year. Has the fatality rate decreased?

Assuming that the populations at risk are the same in the two 12 month periods, $L_0 = L_1$, so Equations (4.11) and (4.12) give an estimated log incidence density ratio of $\ln(50/75) = -0.405$ with standard error 0.183. Consequently a 95% confidence interval for $\ln(IDR)$ is $-0.405 \pm 2 \times 0.183$, which gives the interval estimate (0.46, 0.96) for the incidence density ratio. This suggests that the fatality rate has decreased, but possibly not by very much.

A confidence interval for *IDR* based on *F*-distributions is given by Breslow and Day (1987, page 95), and should be used when the observed counts are small. These authors give a hypothetical illustration of $a = 14$ and $b = 5$ bladder cancer deaths observed among exposed and nonexposed members of an industrial cohort compared with expected numbers based on vital statistics of 5.5 and 7.3, respectively. The estimated incidence density ratio is thus $(14/5.5)/(5/7.3) = 3.72$ and Equation (4.12) gives a standard error of $\sqrt{(1/14+1/5)} = 0.521$, so the approximate 95% confidence interval is $3.72\exp(\pm 2 \times 0.521)$, that is, $(1.31, 10.55)$. This may be compared with the more accurate interval $(1.26, 13.2)$ Breslow and Day give based on the F-distributions.

The stratification methods for odds ratios described in the preceding section carry over to incidence density ratios. The Mantel–Haenszel estimate of the common incidence density ratio based on a stratified cohort study is given by a formula similar to Equation (4.6), as follows, where again g specifies the stratum.

$$IDR_{MH} = \frac{\sum_g a_g L_{0g}/(L_{0g}+L_{1g})}{\sum_g b_g L_{1g}/(L_{0g}+L_{1g})} \qquad (4.13)$$

A formula for the standard error of the logarithm of this estimate is given by Breslow and Day (1987, page 109). This formula is a special case, which arises in the limit as the individual risks approach 0, of the Robins formula (4.7) for the standard error associated with the Mantel–Haenszel stratified odds ratio given by Equation (4.6).

As with odds ratios, the assumption of homogeneity of incidence density ratios based on different strata may be tested using a chi-squared statistic based on Equations (4.4) and (4.5). Similarly, a Mantel–Haenszel test that the common incidence density ratio is 1 may be carried out using a formula similar to that given by Equation (4.8), namely

$$\chi^2_{MHS} = \left(\sum_{g=1}^{G} \frac{a_g - b_g L_{1g}/L_{0g}}{1 + L_{1g}/L_{0g}} \right)^2 / \sum_{g=1}^{G} \frac{(a_g + b_g)L_{1g}/L_{0g}}{(1+L_{1g}/L_{0g})^2} \qquad (4.14)$$

As an illustration of these methods, let us consider the classic Doll–Hill (1966) cohort study of smoking and coronary deaths among British male doctors, cited in Breslow and Day (1987, page 112). The data are listed in Table 4.9 as follows (SM = smoker).

| | Deaths | | Person-years | | |
Age Group	SM	Non-SM	SM	Non-SM	IDR
35-44	32	2	52,407	18,790	5.74
45-54	104	12	43,248	10,673	2.14
55-64	206	28	28,612	5,710	1.47
65-74	186	28	12,663	2,585	1.36
75-84	102	31	5,317	1,462	0.90

TABLE 4.9: Cohort Study of Smoking and Mortality in British Doctors

The age-specific incidence density ratios and 95% confidence intervals, calculated from Equations (4.11) and (4.12), are graphed in Figure 4.3.

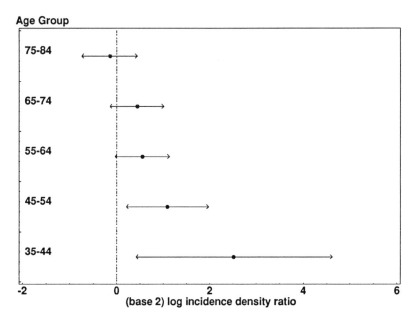

FIGURE 4.3: Results from Doll–Hill Smoking/Mortality Cohort Study

Note that there are only two coronary deaths observed among non-smokers in age group 35–44, so the estimated standard error corresponding to this age group may be inaccurate. Even so, it is clear from the graph that the relative risks decrease with age.

The stratified Mantel–Haenszel estimate of the common incidence density ratio is 1.425, and the 95% confidence interval turns out to be (1.15, 1.77). While this appears to implicate smoking as a risk factor, it is not reasonable to assume a common relative risk. The chi-squared statistic for testing the homogeneity of the incidence density ratios is 10.2 in the present case with 4 degrees of freedom, giving the p-value 0.037.

Even though the null hypothesis of homogeneity should be rejected, the test may not be appropriate here in view of the trend in the estimates of relative risk. If a linear age trend is fitted by weighted least squares to these values, the corresponding chi-squared statistic based on the residuals is 1.40 with 3 degrees of freedom ($p = 0.71$), so the inhomogeneity is entirely explained by a linear trend.

5: Multiple Risk Factors

The examples discussed in the preceding sections all involve a binary exposure variable and a stratification variable which could be a confounder. A stratification variable may be regarded as a risk factor in its own right, and it is then of interest to consider the two risk factors jointly.

In our analysis of the neonatal mortality study we took prenatal care as the exposure of interest and clinic (A or B) as the stratification variable, obtaining an estimate 1.114 for the adjusted odds ratio associated with less than one month of prenatal care. If the exposure of interest is taken to be Clinic B (versus Clinic A) and the level of prenatal care is taken as the stratification variable, the data may be summarised as in Table 4.10, an alternative data layout to Table 4.1.

		PNC < 30 days			PNC 30+ days	
	Clinic:	B	A		B	A
	Yes	17	3	Yes	2	4
Died within 1 month						
	No	197	176	No	23	293
		OR = 5.06			OR = 6.37	

TABLE 4.10: Neonatal Mortality vs Clinic Stratified by Prenatal Care

Equation (4.6) gives a Mantel–Haenszel estimate of 5.27 for the combined odds ratio associated with Clinic B versus Clinic A.

Now consider prenatal care and clinic jointly. In this case the exposure variable has four categories corresponding to the combinations of risk factors, and the data may be presented as follows:

		Exposure Category			
		A:PNC+	A:PNC-	B:PNC+	B:PNC-
	Dead	4	3	2	17
Outcome					
	Alive	293	176	23	197

Even though there are four exposure categories rather than two, you may still proceed with the analysis by choosing one of the exposure levels as the referent exposure, and computing the odds ratio for each other exposure category with respect to it. As you saw in Section 2 of Chapter 3, a separate confidence interval may be computed for each odds ratio.

Choosing the first category (more prenatal care in Clinic A) as

the referent exposure, the following estimates and confidence intervals
for the odds ratios are obtained (Table 4.11).

Clinic/PNC	Odds Ratio	95% Confidence Interval
A: PNC+	(1)	
A: PNC–	1.25	(0.27, 5.82)
B: PNC+	6.37	(1.07, 37.97)
B: PNC–	6.32	(2.05, 19.50)

TABLE 4.11: Neonatal Mortality Odds Ratios

When computing odds ratios for multiple risk factors the choice
of referent group may be important: wider confidence intervals will
generally result if the referent group contains smaller numbers. On the
other hand, it may be desirable to choose the lowest risk group as the
referent category, so that all the odds ratio are greater than 1.

In Table 4.12 these odds ratios are presented in another way,
which highlights the joint effect of the two risk factors. The numbers
in the margins are the (Mantel–Haenszel) odds ratios associated with
each risk factor after adjusting for the other. This table may also be
used to get the odds ratio for each risk factor corresponding to both
levels of the other factor, by dividing the odds ratio in the bottom
corner of the table by the off-diagonal odds ratios. For example, the
odds ratio for Clinic B compared to Clinic A when there is less prenatal care is computed as 6.32/1.25 = 5.06 (in agreement with the
result shown in Table 4.10), and the odds ratio for less prenatal care
compared with more prenatal care in Clinic B is 6.32/6.37 = 0.99 (in
agreement with the result given after Table 4.1).

		Clinic		
		A	B	
	30+ days	(1)	6.37	(1)
Prenatal Care				
	< 30 days	1.25	6.32	1.11
		(1)	5.27	

TABLE 4.12: Odds Ratios for Two Risk Factors: Neonatal Mortality Study

If there is no effect modification the odds ratios in the margins
will be similar to their corresponding off-diagonal odds ratios in the
table, any differences being attributable purely to sampling variation.
Furthermore, in the absence of effect modification the odds ratio in the
bottom right corner of the table should be equal to the product of the

off-diagonal odds ratios: this means that the two risk factors act independently of each other.

Table 4.13 gives similar summaries of the odds ratios for combinations of the two risk factors for each of the four confounding examples from Section 2. In each case the referent group is taken as that with no exposure to each risk factor.

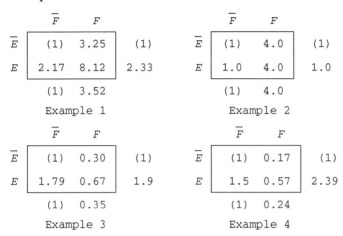

TABLE 4.13: Odds Ratios for Two Risk Factors: Confounding Examples

In Example 1 there is no confounding while in Examples 2 and 3 there is confounding but no effect modification. In Example 4 the situation is more complicated because the odds ratios across strata are different, so the Mantel–Haenszel adjusted odds ratios are not valid measures. Here the odds ratio associated with E is 1.5 in the absence of F and 3.4 (0.57/0.17) when F is present, while the odds ratio associated with F is 0.17 for E absent and 0.38 (0.57/1.5) for E present.

As a more complex illustration of multiple risk factors, involving an additional stratification variable, consider some data collected by Tuyns et al (1977) and cited extensively by Breslow and Day (1980) for a case-control study investigating risk factors for oesophageal cancer. Two risk factors were considered: average daily consumption of (1) tobacco and (2) alcohol, and the data were also classified into 10-year age groups, as shown in Table 4.14.

There are nine exposure groups corresponding to the combinations of the three levels of tobacco consumption with the three alcohol consumption levels. Taking the combination of 0–39 grams per day of alcohol and 0–9 grams per day of tobacco as the baseline exposure level, age-adjusted Mantel–Haenszel odds ratios may be calculated using Equation (4.6) for each of the remaining eight exposure levels. For example, the calculations for the combination

(Alcohol: 40–79, Tobacco: 0–9) are as follows.

$$OR_{MH} = \frac{0/67 + 0/95 + (6 \times 45)/84 + (9 \times 47)/89 + (17 \times 43)/82 + (2 \times 17)/23}{0/67 + 0/95 + 32/84 + (2 \times 31)/89 + (5 \times 17)/82 + 3/23}$$

$$= 8.18$$

Age		Tobacco: 0-9 0-39	40-79	80+	Tobacco 10-19 0-39	40-79	80+	Tobacco: 20+ 0-39	40-79	80+
25-34	Case	0	0	0	0	0	1	0	0	0
	Cont.	40	27	3	10	7	1	11	11	5
35-44	Case	0	0	2	1	3	0	0	1	2
	Cont.	60	35	12	13	20	9	15	21	5
45-54	Case	1	6	7	0	4	9	0	10	9
	Cont.	45	32	13	18	17	9	14	12	7
55-64	Case	2	9	14	3	6	14	7	7	14
	Cont.	47	31	14	19	15	8	11	16	5
65-74	Case	5	17	9	4	3	5	2	5	5
	Cont.	43	17	8	10	7	9	7	4	1
75+	Case	1	2	3	2	1	2	1	1	0
	Cont.	17	3	0	4	2	0	2	3	0

TABLE 4.14: Data from Oesophageal Cancer Case-Control Study

The age-adjusted odds ratios calculated in this way for the various exposure groups are given in Table 4.15 and graphed in Figure 4.4. The values in the margins of Table 4.15 are obtained by adjusting with respect to both age and the other exposure factor. For example the value 3.76 is estimated by comparing the subjects consuming 40–79 grams/day of alcohol with those consuming 0–39 grams/day in the 18 strata corresponding to the six age groups and three tobacco exposure levels, and then using the Mantel–Haenszel formula.

Alcohol Consumption	Tobacco Consumption 0-9 g/day	10-19 g/day	20+ g/day	
0-39 g/day	(1.0)	3.90	5.69	(1.0)
40-79 g/day	8.18	8.63	14.86	3.76
80+ g/day	19.83	19.89	62.35	11.06
	(1.0)	1.51	2.73	

TABLE 4.15: Multiple Odds Ratios from Oesophageal Cancer Study

You can see from Table 4.15 that there is some evidence of effect modification, since the effect of each risk factor depends on the level of the other. For example the odds ratio showing the effect of 40–79 grams/day of alcohol compared to 0–39 grams/day is 8.18 at the lowest level of tobacco consumption, but drops to 2.21 (8.63/3.90) and 2.61 (14.86/5.69) at the higher levels of tobacco consumption. This means that the effects of alcohol and tobacco on the disease may not be independent.

The graph in Figure 4.4 shows the base 2 logarithms of the adjusted odds ratios and their corresponding confidence intervals, calculated from Equation (4.7). (Base 2 logarithms are taken so that each unit on the horizontal scale corresponds to a doubling of the odds ratio, facilitating interpretation.)

Figure 4.4 is similar to a two-way anova plot (described in Section 4 of Chapter 3). If there is no effect modification the effects of the two factors are multiplicative, or additive on a logarithmic scale, and you can see from Figure 4.4 that this is not the case. However, the confidence intervals are quite wide and could easily accommodate an independent effects model, so the evidence of effect modification is not strong.

It is also interesting to note that the effect modification would disappear if the odds of disease in two low-alcohol cells were smaller by a factor of 4 relative to the referent group.

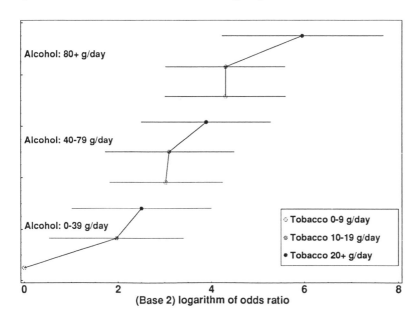

FIGURE 4.4: Odds Ratios for Multiple Exposures: Oesophageal Cancer

Summary

This chapter has been devoted to methods for analysing data presented as sets of 2-by-2 contingency tables. These methods are known in the epidemiology literature as Mantel–Haenszel methods. Measures of association include the risk difference, the relative risk, and the odds ratio. The odds ratio is the preferred measure of association in a 2-by-2 table of counts because it is valid for any study type or level of risk.

Confounding

Confounding is a distortion of an odds ratio that occurs when a second risk factor is associated with the risk factor of interest. Four hypothetical examples were introduced to gain an understanding of how confounding works.

Combining Odds Ratios

In this section you saw how to adjust an odds ratio for the effect of confounding by a stratification variable, using either a generally valid asymptotic formula or the more robust Mantel–Haenszel method. The method was applied to a study of possible sex discrimination in admission to graduate programs at Berkeley in 1973.

The Relative Risk

The relative risk may be measured for a cohort or cross-sectional study. For a cohort study where different subjects could have different durations of exposure, the incidence density ratio is used to measure the relative risk, and provided the individual risks are small the methods for analysing odds ratios apply. You saw an application of these methods to a classic cohort study linking smoking to coronary deaths among British doctors.

Multiple Risk Factors

A case-control study in a French village investigating smoking and alcohol consumption as risk factors for oesophageal cancer was used to illustrate the application of Mantel–Haenszel methods to multiple risk factors. You saw how odds ratios for the various exposure combinations compared to a low exposure referent group could be displayed in an informative summary table and graphed with confidence intervals using a display similar to an anova plot.

Exercises

Exercise 4.1: Some data from a cohort study of OC use in relation to the development of bacteriuria reported by Evans et al (1978) are given in the following table. Compute the risks of bacteriuria in the exposed and nonexposed groups, and thus the increase in the risk due to OC use. Also give a 95% confidence interval for the relative risk.

		OC Use	
		Yes	No
	Yes	27	75
Bacteriuria			
	No	455	1831

Exercise 4.2: Data from a cohort study examining a possible association between postmenopausal hormone use and coronary heart disease among nurses reported by Stampfer et al (1985) are as follows.

	Postmenopausal Hormone Use	
	Yes	No
Coronary Heart Disease	30	60
Person-years	54308	51477

Estimate the relative risk of disease for hormone use, and give a 95% confidence interval and a *p*-value for testing the association.

Exercise 4.3: The following data arose from an experiment investigating motion sickness at sea reported by Burns (1984), and are also discussed by Altman (1991, pages 368–371). The study subjected individuals to vertical motion for two hours, recording the time until each subject vomited or asked to withdraw. Those in Group 1 were given frequency 0.167 Hz and acceleration 0.111 g, while those in Group 2 were subjected to a double dose (0.333 Hz and 0.222 g). An asterisk denotes withdrawal or 'successful' completion.

Group 1:   30 50 50* 51 66* 82 92 120* 120* 120* 120* 120* 120*
             120* 120* 120* 120* 120* 120* 120* 120*

Group 2:   5 6* 11 11 13 24 63 65 69 69 79 82 82 102 115 120* 120*
             120* 120* 120* 120* 120* 120* 120* 120* 120* 120* 120*

Estimate the incidence density ratio for occurrence of an adverse event in Group 2 compared to Group 1, and give a 95% confidence interval.

Exercise 4.4: The following data are taken from a case-control study reported by Buring et al (1983) to examine the association between exercise and myocardial infarct, stratified by cigarette smoking history.

Smoking history	Physical activity index			
	2500+ kcals		<2500 kcals	
	cases	controls	cases	controls
Never smoked	41	84	46	52
Exsmoker 10+ yrs	41	80	30	39
Exsmoker <10 yrs	22	34	21	26
Current smoker	86	68	79	40

Estimate the odds ratios within each strata, and test the hypothesis that they do not vary between strata using a goodness-of-fit test. Compute the common odds ratio and give a 95% confidence interval for it. Is smoking a confounding variable? Explain.

Exercise 4.5: The following table gives data, stratified by age group and smoking level, from a case-control study reported by Shapiro et al (1979) investigating oral contraceptive (OC) use as a risk factor for myocardial infarction among women. Compute the age-stratified Mantel–Haenszel odds ratios for each combination of the two risk factors OC use and smoking level, using non-smoking OC users as the referent group, and present the results in a table. Include in the margins of this table the Mantel–Haenszel odds ratio estimates for each risk factor adjusted for the other (as well as age).

	Cigs/day:	0		1-24		25+	
Age		OC	No OC	OC	No OC	OC	No OC
25-29	Case	1	0	0	1	1	3
	Cont.	107	25	79	26	40	15
30-34	Case	0	0	5	1	7	8
	Cont.	175	13	147	11	80	18
35-39	Case	3	0	11	1	19	3
	Cont.	156	8	130	12	77	10
40-44	Case	10	1	21	0	34	5
	Cont.	175	5	151	4	101	6
45-49	Case	20	3	42	0	21	3
	Cont.	175	5	138	1	81	5

Exercise 4.6: In an experiment to investigate the effect of cutting length and planting time on the growth prospects of plum root cut-

tings, 240 cuttings were plant for each of four combinations of these risk factors, with the following results (see Bartlett, 1935). Analyse these data by comparing odds ratios.

Cutting length	Planting time	No. survived
long	at once	156
long	in Spring	84
short	at once	107
short	in Spring	31

References

Altman, D.G (1991): *Practical Statistics for Medical Research.* Chapman & Hall. London.

Bartlett, M.S. (1935): Contingency table interactions, *Journal of the Royal Statistical Society Supplement*, **2**, pages 248–252.

Bishop, Y.M.M. (1969): Full contingency tables, logits, and split contingency tables, *Biometrics*, **25**, pages 383–399.

Breslow, N.E. & N.E. Day (1980): *Statistical Methods in Cancer Research: Volume I – The Analysis of Case-Control Studies.* IARC. Lyon.

Breslow, N.E. & N.E. Day (1987): *Statistical Methods in Cancer Research: Volume II – The Design and Analysis of Cohort Studies.* IARC. Lyon.

Buring, J.E. et al (1983): Alcohol and HDL in non-fatal myocardial infarction: preliminary results from a case-control study, *Circulation*, **68**, page 227.

Burns, K.C. (1984): Motion sickness incidence: distribution of time to first emesis and comparison of some complex motion sickness conditions, *Aviation Space and Environmental Medicine*, **56**, pages 521–527.

Doll, R. & A.B. Hill (1966): Mortality of British doctors in relation to smoking: observations on coronary thrombosis. *Natl Cancer Inst. Monogr.*, **19**, pages 205–268.

Evans, D.A., C.H. Hennekens, L. Miao, L.W. Laughlin, W.G. Chapman, B. Rosner, J.O. Taylor, & E.H. Kass (1978): Oral-contraceptive use and bacteriuria in a community-based study, *N.Engl.J.Med.* **299**, pages 536–537.

Freedman, D., R. Pisani & R. Purves (1978): *Statistics.* W.W. Norton & Co. New York.

Kleinbaum, D.G., L.L. Kupper & H. Morgenstern (1982): *Epidemiologic Research.* Van Nostrand Reinhold. New York.

Mantel, N. & W. Haenszel (1959): Statistical aspects of the analysis of data from retrospective studies of disease, *Journal of the National Cancer Institute*, **22**, pages 719–748.

Meier, P. (1972): The biggest public health experiment ever: the 1954 field trial of the Salk poliomyelitis Vaccine, in *Statistics: A Guide to the Unknown* (edited by J.M. Tanur), pages 2–13. Holden-Day. San Francisco.

Robins, J.M., N.E. Breslow & S. Greenland (1986): Estimators of the Mantel-Haenszel variance consistent in both sparse data and large-strata limiting models, *Biometrics*, **92**, pages 311–323.

Shapiro, S, L. Rosenberg, D. Slone & D.W. Kaufman (1979): Oral-contraceptive use in relation to myocardial infarction, *The Lancet*, 7 April, pages 743–747.

Stampfer, M.A., W.C. Willett, G.A. Colditz, B. Rosner, F.E. Speizer, & C.H. Hennekens: (1985): A prospective study of post-menopausal estrogen therapy and coronary heart disease, *N.Engl.J. Med.* **313**, pages 1044–1049.

Tuyns, A.J., G. Péquignot & O.M. Jenson (1977): Le cancer de l'oesophage en Ille-et-Vilaine en fonction des niveaux de consommation d'alcool et de tabac. *Bull.Cancer* **64**, pages 45–60.

5
LOGISTIC REGRESSION I

1: Introduction

In the preceding chapter some basic statistical methods for analysing data from epidemiological study designs were introduced. These methods involve the analysis of two-by-two contingency tables relating a dichotomous outcome to a dichotomous exposure. We focused on the odds ratio as the preferred measure of association and gave formulas enabling confidence intervals to be computed. You saw how, if a third variable (a possible confounder) needs to be considered, a stratified analysis may be carried out.

A problem with stratified analysis is that if the data are too thinly spread over the strata, the estimates for these strata, not to mention their standard errors, are rather imprecise. While the Mantel–Haenszel estimates of a common effect are still robust, you need to check that you can justifiably combine the components from different strata. It is reasonable to do this if the stratification variable is simply a confounder, but not if the effect varies across strata (as is the case when the stratification variable is an effect modifier rather than a simple confounder).

In this chapter a statistical model for analysing epidemiological data is introduced. This model, the *logistic regression* model, is similar to the linear regression model used in much of classical statistical methodology. The main difference is that whereas in linear regression you assume a continuous response or outcome variable, the logistic regression model requires that the response variable be dichotomous. Thus the logistic model is well suited to the analysis of the binary outcome data considered in Chapter 4.

Another advantage of using the logistic regression model to analyse data from epidemiological studies is that it can handle quite general exposure variables, not just dichotomous ones.

For a single exposure variable E, the model takes the form

$$\ln\left(\frac{P}{1-P}\right) = a + b\,x \tag{5.1}$$

where P denotes the probability of occurrence of the outcome D and x is the value of an exposure E. This equation may be inverted to give an expression for the probability P as

$$Prob[D] = \frac{1}{1 + \exp(-a - bx)} \qquad (5.2)$$

The functional form of the right-hand side of Equation (5.2) ensures that its values are always between 0 and 1, which is reasonable given that they are probabilities. The function relating these probabilities to the values of x is shown in Figure 5.1, where it is assumed that the parameter b is positive.

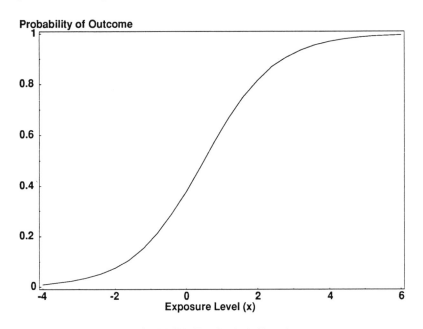

FIGURE 5.1: The Logistic Function

Using the logistic model it is a straightforward matter to derive formulas for the various measures of association arising from a two-by-two contingency table. In this case the exposure has just two possible values of x which may be taken to be 0 (nonexposed) and 1 (exposed). The risk of the outcome given the exposure is thus obtained by putting $x = 1$ in Equation (5.2), that is

$$Prob[D \mid E] = \frac{1}{1 + \exp(-a - b)} \qquad (5.3)$$

while the risk of the outcome given no exposure ($x = 0$) is

$$Prob[D \mid \bar{E}] = \frac{1}{1 + \exp(-a)} \qquad (5.4)$$

The relative risk is the ratio of these two expressions. However, the expressions for the odds and the odds ratio are much simpler. The odds of the outcome given the exposure is, from Equation (5.3),

$$\frac{Prob[D \mid E]}{Prob[\bar{D} \mid E]} = \frac{Prob[D \mid E]}{1 - Prob[D \mid E]} = \frac{\dfrac{1}{1 + \exp(-a-b)}}{1 - \dfrac{1}{1 + \exp(-a-b)}}$$

and this reduces after cancellation to $\exp(a+b)$. Actually this result may be obtained more easily by recognising that the left-hand side of Equation (5.1) is just the natural logarithm of the odds, and substituting $x = 1$ gives $a + b$ for the log odds, or $\exp(a+b)$ for the odds.

Similarly, the odds of the outcome given no exposure is

$$\frac{Prob[D \mid \bar{E}]}{1 - Prob[D \mid \bar{E}]} = \frac{\dfrac{1}{1 + \exp(-a)}}{1 - \dfrac{1}{1 + \exp(-a)}} = \exp(a)$$

which is also obtained by putting $x = 1$ in Equation (5.1). Thus the odds ratio has the simple formula

$$OR = \frac{\exp(a+b)}{\exp(a)} = \exp(b) \qquad (5.5)$$

The log of the odds ratio based on the logistic model is even simpler: you can see from Equation (5.5) that it is just b. This means that the parameter b in the model may be interpreted directly as the (natural) logarithm of the odds ratio.

The logistic model given by Equations (5.1) and (5.2) is easily fitted to data from a two-by-two contingency table. As an illustration, recall the data from the study relating prenatal care to neonatal mortality discussed in Section 2 of Chapter 4.

		Exposure Prenatal Care (days)	
		0-29	30+
Outcome	Died	20	6
	Survived	373	316

Some printout from a statistical package obtained as a result of fitting the logistic regression model to these data is given in Table 5.1.

```
Logistic Regression Analysis
Response: Died within 1 month
  Col Name        Coeff   StErr p-value  Odds  95%     CI

    0 Constant -3.964    0.412   0.000
    3 Less PNC  1.038    0.472   0.028 2.824 1.12 7.118

df:0              Dev:0      %(0):96.364 #iter:9    RSq:1
```

TABLE 5.1: Printout from Simple Logistic Model: Neonatal Mortality DATA

The most important numbers in this printout are those in the right-hand side entitled Odds and 95% CI: these are the estimated odds ratio (2.824) and its 95% confidence interval (1.12, 7.118). In fact these results correspond to those given by Equations (4.1) and (4.2). The printout in Table 5.1 gives an asymptotically precise confidence interval having half-width equal to 1.96 times the standard error.

It is also useful to examine the *p*-value for the exposure variable (0.028). This is based on using the odds ratio as a test statistic for the null hypothesis that the odds ratio in the target population is 1.

The numbers labelled Coeff and StErr are the estimate of the parameter *b* (the natural logarithm of the odds ratio) and its standard error, respectively.

The row in the printout commencing with 0 Constant gives the value of the other parameter, *a*, and its associated standard error and *p*-value. Unlike the parameter *b*, *a* is simply a normalising constant and does not have an interpretation as an odds ratio.

The numbers in the bottom line of the printout are as follows. The first (df) is the number of degrees of freedom remaining in the data table after fitting the logistic model, and is always 0 for a two-by-two table, since there are two independent pieces of data (the odds of the outcome in each of the two exposure categories) to be fitted and there are two parameters (*a* and *b*) in the model. The second number (Dev) is the *deviance*, a measure of the error after fitting the model (similar to the residual sum of squares in ordinary linear regression). Again this must be 0 for a two-by-two table since the model is a perfect fit and there is thus no error associated with it. For the same reason the last number on the bottom line (RSq, which stands for *r-squared*, a measure of goodness-of-fit necessarily between 0 and 1) is always equal to 1 for a two-by-two table. Finally, %(0) is the overall percentage of non-outcomes (useful for data-checking purposes) and #iter is the number of steps in the model fitting procedure.

The logistic model may be used to fit data from any two-by-two contingency table, no matter whether the data arise from a cross-sectional, cohort, or case-control study. Of course if the data are so simple there is no benefit in fitting the model, since all the relevant statistics are given by simple formulas. The usefulness of the logistic model arises from its ability to handle covariates.

If there are p predictor variables x_1, x_2, ..., x_p, the logistic regression model takes the general form

$$Prob[D] = \frac{1}{1 + \exp(-a - \sum_{j=1}^{p} b_j x_j)} \qquad (5.6)$$

generalising Equation (5.2). If one of the predictor variables, say x_1, is a dichotomous exposure variable, the argument leading to Equation (5.5) goes through unchanged, since the contribution to the odds from the other variables still cancels out.

More generally, the general logistic regression model may be used to estimate the odds ratio for any combination of predictor variable compared to any other combination. To see how this works, suppose that the two combinations are denoted by E_1 and E_0, respectively, where

$$E_k = (x_1^{(k)}, x_2^{(k)}, \ldots, x_p^{(k)})$$

Using Equation (5.5) the odds of the outcome D, given that the values of the predictor are equal to those in combination E_k, is

$$\frac{Prob[D \mid E_k]}{1 - Prob[D \mid E_k]} = \exp(a + \sum_{j=1}^{p} b_j x_j^{(k)})$$

The odds ratio for combination E_1 to E_0 is thus

$$OR = \exp\left(\sum_{j=1}^{p} b_j (x_j^{(1)} - x_j^{(0)}) \right) \qquad (5.7)$$

so the log of the odds ratio is just a linear combination of the parameters in the model. Note that Equation (5.5) is the special case of Equation (5.7) which arises when $p = 1$ and $x_1^{(1)} = 1$, $x_1^{(0)} = 0$.

To fit the general logistic model using a statistical package, the data layout needs to follow a certain structure similar to that used in multiple regression analysis. The details are given in the next section, where logistic regression is also used to analyse the confounding examples from the preceding chapter.

2: Modelling Confounding

The usual data layout for fitting the general logistic regression model takes the form of an array with each row corresponding to a particular combination of the predictor variables. There is a column in the array corresponding to each predictor variable, and two further columns are required to specify (1) the number of outcomes, and (2) the total number of individuals, corresponding to the various combinations of predictors. If there is a single dichotomous predictor for exposure level, the array will have just two rows and three columns, as follows.

D	$Total$	E
y_1	n_1	1
y_0	n_0	0

Here y_1 and y_0 are the numbers having the outcome out of totals n_1 in the exposed group and n_0 in the nonexposed group. Thus for the study relating prenatal care and neonatal mortality, the data layout is as shown next.

D	$Total$	E
20	393	1
6	322	0

If there are additional covariates the data array is more complex. Consider the inclusion of prenatal care clinic as a covariate F, taking the values 1 for Clinic A and 0 for Clinic B. The data layout for the logistic regression analysis is as follows.

	Prenatal care < 30 days					
	Yes	No		Yes	No
Died within 1 month — Yes	3	4	Yes	17	2
Died within 1 month — No	176	293	No	197	23
	Clinic A			Clinic B		

D	$Total$	E	F
3	179	1	1
4	297	0	1
17	214	1	0
2	25	0	0

To fit a logistic regression model to these data, both E and F are included as predictor variables, and the resulting printout is shown in Table 5.2.

```
Logistic Regression Analysis
Response: Died within 1 month
Col Name        Coeff  StErr p-value  Odds   95%    CI

  0 Constant   -2.549  0.561  0.000
  3 Less PNC    0.110  0.561  0.844  1.117 0.372 3.353
  4 Clinic A   -1.699  0.531  0.001  0.183 0.065 0.517

df:1    Dev:0.043   %(0):96.364   #iter:9   RSq: 0.998
```

TABLE 5.2: Multiple Logistic Model Printout: Neonatal Mortality Data

This printout shows an odds ratio of 1.117 for the effect of less prenatal care on neonatal mortality after adjusting for the effect of the clinic. This estimate may be compared with the Mantel–Haenszel estimate calculated from Equation (4.6), which is 1.114. Similarly the printout shows a 95% confidence interval (0.372, 3.353) for the odds ratio, which is comparable with the interval (0.376, 3.330) calculated from the formula (4.7) using 1.96 times the standard error as the half-width of the confidence interval.

The printout in Table 5.2 also gives an estimated odds ratio and confidence interval for the effect of Clinic A compared with Clinic B: the conclusion is that there is a benefit associated with Clinic A.

The bottom line of the printout shows that there is 1 degree of freedom, a consequence of fitting a model with 3 parameters to 4 observations (the odds of outcome for each exposure combination). The deviance provides a measure of how well the logistic model fits the data. A p-value for this goodness-of-fit test is given by area in the upper tail of a chi-squared distribution with the given number of degrees of freedom. In the present case the deviance is 0.043, corresponding to a p-value 0.84, so the fit is good. The r-squared statistic is another measure of the goodness-of-fit taking values between 0 (null fit) to 1 (perfect fit), and is thus analogous to the r-squared statistic arising in linear regression analysis. In the present case its value is 0.998, indicating an almost perfect fit.

As in ordinary linear regression analysis, a logistic regression model may be reduced by omitting a predictor variable. Table 5.3 shows what happens to the printout when a reduced logistic model containing only clinic as a predictor is fitted to the same data.

This result is almost the same as would be obtained by fitting the logistic model to a single two-by-two table classifying the neonatal mortality outcome by clinic. The only difference is in the bottom line of the printout. Even though the model does not contain a parameter for prenatal care, disaggregating the data by level of prenatal care provides 2 degrees of freedom, from which the goodness-of-fit of the model may be assessed.

```
Logistic Regression Analysis
Response: Died within 1 month
 Col Name          Coeff  StErr p-value  Odds    95%    CI
```

Col	Name	Coeff	StErr	p-value	Odds	95%	CI
0	Constant	-2.449	0.239		0		
4	Clinic A	-1.755	0.450		0	0.173 0.072	0.417

```
df:2     Dev:0.082    %(0):96.364    #iter:8   RSq: 0.995
```

TABLE 5.3: Reduced Logistic Model: Neonatal Mortality Data

Now consider the logistic modelling approach to the confounding examples introduced in Section 3 of Chapter 4.

Example 1

The data layout for analysis, and the printout from fitting the logistic model with both predictors, are as follows.

D	$Total$	E	F
50	90	1	1
30	90	0	1
30	120	1	0
20	150	0	0

```
Logistic Regression Analysis
Response: D
 Col Name          Coeff  StErr p-value  Odds    95%    CI
```

Col	Name	Coeff	StErr	p-value	Odds	95%	CI
0	Constant	-1.914	0.204		0		
3	E	0.848	0.222		0	2.334 1.511	3.607
4	F	1.257	0.221		0	3.516 2.279	5.422

```
df:1     Dev:0.104    %(0):71.111    #iter:9   RSq:0.998
```

The printout indicates that both E and F are significant risk factors and the model fits the data almost perfectly, with R-squared 0.998.

The next printouts show what happens when each predictor variable is omitted in turn.

The unadjusted odds ratios for each risk factor are very close to the adjusted odds ratios, indicating little or no confounding. However you can see from the large deviances that the fit is unsatisfactory unless both predictors are included in the model.

```
Logistic Regression Analyses
Response: D
Col Name          Coeff   StErr p-value  Odds    95%    CI
```

Col	Name	Coeff	StErr	p-value	Odds	95%	CI
0	Constant	-1.335	0.159	0			
3	E	0.849	0.213	0	2.338	1.540	3.551

```
df:2      Dev:33.749  %(0):71.111  #iter:8   RSq:0.326
```

Col	Name	Coeff	StErr	p-value	Odds	95%	CI
0	Constant	-1.482	0.157	0			
3	F	1.258	0.217	0	3.52	2.301	5.384

```
df:2      Dev:15.067  %(0):71.111  #iter:8   RSq:0.699
```

The estimated odds ratios corresponding to each exposure combination may be calculated from the fitted model using Equation (5.7). The model is the special case of Equation (5.6) where there are two predictor variables x_1 and x_2, that is, substituting the estimates given in the first printout,

$$OR = \exp\left(0.848(x_1^{(1)} - x_1^{(0)}) + 1.257(x_2^{(1)} - x_2^{(0)})\right)$$

Choosing the combination $x_1^{(0)} = 0$ and $x_2^{(0)} = 0$ as the referent category, the odds ratios for the other three combinations are obtained from the above formula as follows.

$x_1^{(1)} = 0, x_2^{(1)} = 1:$ $\exp(0.848 \times 0 + 1.257 \times 1) = 3.52$

$x_1^{(1)} = 1, x_2^{(1)} = 0:$ $\exp(0.848 \times 1 + 1.257 \times 0) = 2.34$

$x_1^{(1)} = 1, x_2^{(1)} = 1:$ $\exp(0.848 \times 1 + 1.257 \times 1) = 8.21$

Note that the odds ratio for the combination $x_1^{(1)} = 1$, $x_2^{(1)} = 1$ is equal to the product of the other two odds ratios: that is, the model assumes that the two risk factors act independently. The odds ratios obtained from fitting the logistic model are usefully displayed in a table similar to Table 4.2. Before doing this let us consider the other examples.

Example 2

Recall that Example 2 provides an illustration of confounding where the crude odds ratio for the risk factor E is 2.1 and the adjusted odds ratio is 1. First, look at the printout obtained from fitting the model that includes both predictor variables. In agreement with the Mantel–Haenszel analysis for these data given in Section 3 of Chapter 4, this printout shows that E is not a risk factor.

```
Logistic Regression Analysis
Response: D
Col Name          Coeff    StErr p-value  Odds   95%     CI

   0 Constant   -1.386    0.189     0
   3 E           0.000    0.267     1 1.000 0.592 1.689
   4 F           1.386    0.270     0 3.999 2.356 6.787

   df:1              Dev:0      %(0):65      #iter:8       RSq: 1
```

Now look at what happens when *F* and *E* are omitted from the analysis in turn.

```
Logistic Regression Analyses
Response: D
Col Name          Coeff    StErr p-value  Odds   95%     CI

   0 Constant   -1.030    0.165   0.000
   3 E           0.742    0.216   0.001    2.1 1.376 3.206

   df:2          Dev:28.409    %(0):65      #iter:8   RSq: 0.299

Col Name          Coeff    StErr p-value  Odds   95%     CI

   0 Constant   -1.386    0.177     0
   4 F           1.386    0.226     0 3.999 2.566 6.233

   df:2              Dev:0      %(0):65      #iter:8       RSq: 1
```

These results indicate that *F* is a risk factor, with an estimated odds ratio 4, but *E* is not. The odds ratio associated with *E* changes from 2.1 to 1 when an adjustment is made for *F*, so *F* is a confounder for *E*. However, the odds ratio associated with *F* is unchanged by adjusting for *E*, so *E* is not a confounder for *F*. This shows that confounding is not a symmetric phenomenon.

Example 3

As you saw in the discussion in Chapter 4, Example 3 is a further illustration of confounding, where the crude odds ratio for the risk factor *E* is 1 and the adjusted odds ratio is close to 2. The relevant printouts are as follows.

```
Logistic Regression Analyses
Response: D
Col Name          Coeff    StErr p-value  Odds   95%     CI

   0 Constant    0.098    0.102   0.339
   3 E           0.640    0.191   0.001 1.897 1.305 2.756
   4 F          -1.043    0.195   0.000 0.352 0.241 0.516

   df:1          Dev:0.271    %(0):50      #iter:8   RSq: 0.991
```

Col	Name	Coeff	StErr	p-value	Odds	95%	CI
0	Constant	0	0.100	1			
3	E	0	0.141	1	1	0.758	1.319

df:2	Dev:30.902	%(0):50	#iter:5	RSq: 0

Col	Name	Coeff	StErr	p-value	Odds	95%	CI
0	Constant	0.251	0.092	0.006			
4	F	-0.631	0.146	0.000	0.532	0.399	0.709

df:2	Dev:12.066	%(0):50	#iter:8	RSq: 0.61

These results indicate that both *E* and *F* are risk factors, and each is a confounder for the other. Leaving out *F* completely masks the association between *E* and the outcome *D*, and omitting *E* substantially reduces the association between *F* and *D*.

Example 4

The following printout is obtained when the logistic regression model with both risk factors is fitted to the data from Example 4.

```
Logistic Regression Analysis
Response: D
```

Col	Name	Coeff	StErr	p-value	Odds	95%	CI
0	Constant	0.335	0.112	0.003			
3	E	0.880	0.199	0.000	2.412	1.633	3.563
4	F	-1.426	0.200	0.000	0.240	0.162	0.355

df:1	Dev:4.287	%(0):50	#iter:9	RSq: 0.931

This result is similar to that for Example 3, but a deviance of 4.287 with 1 degree of freedom gives a *p*-value 0.038 for the homogeneity test, so the fit is inadequate. Recall that the estimated odds ratios for the association between *E* and *D* are 3.4 when *F* is present and 1.5 when *F* is absent, and it is not reasonable to fit a common odds ratio for this association. This is an example of effect modification rather than simple confounding.

Multiple Risk Factor Tables

The odds ratios obtained from the model-fitting may now be displayed in Table 5.4, using a similar format to that given in Table 4.3.

The numbers in the margins are the adjusted odds ratios associated with each of the two risk factors, and these are necessarily the same as the odds ratios in the body of the table because the model assumes no effect modification. You may confirm this fact by using Equation (5.7) to check that (a) the odds ratio for comparing the exposure $(x_1 = 1, x_2 = 1)$ versus $(x_1 = 0, x_2 = 1)$ is equal to the odds ratio for comparing $(x_1 = 1, x_2 = 0)$ versus $(x_1 = 0, x_2 = 0)$, and that (b) the odds ratio for $(x_1 = 1, x_2 = 1)$ versus $(x_1 = 1, x_2 = 0)$ is equal to the odds ratio for $(x_1 = 0, x_2 = 1)$ versus $(x_1 = 0, x_2 = 0)$. In other words, the odds ratio associated with E does not depend on F, and vice versa.

	$\bar{F}$	F	
$\bar{E}$	(1) 3.52	(1)
E	2.33 8.21	2.33
	(1) 3.52	

Example 1

	$\bar{F}$	F	
$\bar{E}$	(1) 4.0	(1)
E	1.0 4.0	1.0
	(1) 4.0	

Example 2

	$\bar{F}$	F	
$\bar{E}$	(1) 0.35	(1)
E	1.90 0.67	1.90
	(1) 0.35	

Example 3

	$\bar{F}$	F	
$\bar{E}$	(1) 0.24	(1)
E	2.41 0.58	2.41
	(1) 0.24	

Example 4

TABLE 5.4: Odds Ratios from Logistic Models: Confounding Examples

The results shown in Table 5.4 should be compared with the corresponding results obtained from the Mantel–Haenszel analysis, given in Table 4.3. The agreement between the two approaches is perfect for Example 2 and good for Examples 1 and 3, but not so good for Example 4 where effect modification exists.

The logistic model can handle effect modification by including an *interaction term* – a product of the predictor variables – in the model. The data layout needs to have an extra column for the product, as follows:

D	*Total*	E	F	$E \times F$
155	335	1	1	1
20	100	0	1	0
45	65	1	0	0
180	300	0	0	0

The following printout is now obtained when all three predictors are included in the model. The interaction term is statistically significant ($p = 0.038$) so it should be retained in the model.

```
Logistic Regression Analysis
Response: D
  Col Name         Coeff   StErr p-value  Odds   95%    CI
```

Col Name	Coeff	StErr	p-value	Odds	95%	CI
0 Constant	0.405	0.118	0.001			
3 E	0.406	0.293	0.167	1.500	0.844	2.666
4 F	-1.792	0.276	0.000	0.167	0.097	0.287
5 E*F	0.831	0.401	0.038	2.296	1.047	5.036

```
df:0           Dev:0    %(0):50    #iter:9      RSq: 1
```

To complete the analysis, let us use Equation (5.7) to calculate the odds ratios corresponding to the two risk factors and display them in a table. In this case the model is the special case of Equation (5.6) where there are three predictor variables x_1, x_2, and $x_3 = x_1 x_2$, so substituting the estimates given in the above printout, you get

$$OR = \exp\left(0.406(x_1^{(1)} - x_1^{(0)}) - 1.792(x_2^{(1)} - x_2^{(0)}) + 0.831(x_1^{(1)}x_2^{(1)} - x_1^{(0)}x_2^{(0)})\right)$$

Again choosing the combination $x_1 = 0$ and $x_2 = 0$ as the referent category, the odds ratios for the other three combinations are obtained from the above formula and tabulated as follows:

$x_1^{(1)} = 0, x_2^{(1)} = 1$: $\exp(0.406 \times 0 - 1.792 \times 1 + 0.831 \times 0) = 0.167$
$x_1^{(1)} = 1, x_2^{(1)} = 0$: $\exp(0.406 \times 1 - 1.792 \times 0 + 0.831 \times 0) = 1.500$
$x_1^{(1)} = 1, x_2^{(1)} = 1$: $\exp(0.406 \times 1 - 1.792 \times 1 + 0.831 \times 1) = 0.574$

	$\overline{F}$	F
$\overline{E}$	(1)	0.17
E	1.5	0.57

In this case the odds ratio corresponding to the exposure $x_1 = 1$ and $x_2 = 1$ is not equal to the product of the other two odds ratios, so the risk factors do not act independently. Since the odds ratio for the effect of E depends on F and vice versa, there is no common odds ratio for either risk factor, and no entries are given in the margins of the table.

Confidence Intervals for Odds Ratios

In Chapter 4 you saw how to calculate a confidence interval for an odds ratio using either Equation (4.2) (if the data arise from a simple 2-by-2 table) or Equation (4.7) (if the Mantel–Haenszel estimate based on Equation (4.6) is used to adjust for a stratification variable).

A confidence interval for an odds ratio may also be found in the process of fitting a logistic model. For odds ratios corresponding to the parameters b_1, b_2, ..., b_p in the model, confidence intervals are shown in the printout. However, there are other odds ratios of interest, such as those corresponding to combinations of risk factors, that are not shown in the printout. While these odds ratio estimates may be calculated directly from the printout using Equation (5.7), further analysis is usually needed to obtain the corresponding standard errors.

As an illustration, suppose you wish to estimate the odds ratio for the combination (less prenatal care, Clinic B) compared with (more prenatal care, Clinic A) for the neonatal mortality study. Refer to the discussion in the preceding chapter leading to Table 4.2. The relevant counts are as follows:

Exposure Category

		B: less PNC	A: more PNC
	Dead	17	4
Outcome			
	Alive	197	293

Using Equations (4.1) and (4.2), the estimated odds ratio is 6.32 and the standard error of its logarithm is 0.563, giving a 95% confidence interval (2.10, 19.1) based on the 1.96 standard error formula.

The printout in Table 5.2 shows the logistic regression analysis of the complete set of data with the additional two exposure categories included. However, this printout gives estimated odds ratios and confidence intervals for the two risk factors separately, whereas a confidence interval for their combined effect is needed.

Assuming that the risk factors act independently, the odds ratio for their combined effect is equal to the product of their individual odds ratios, and may be calculated from Table 5.2. This printout gives 1.117 for the effect of less PNC and 0.183 for the effect of Clinic A (or $1/0.183 = 5.46$ for the effect of Clinic B). Multiplying these odds ratios gives a model-based estimate $1.117 \times 5.46 = 6.10$ for the combined effect of less PNC and Clinic B.

This method provides an estimate of the odds ratio for any combination of risk factors. However to get a corresponding confidence interval you need to recode the data before refitting a logistic model. To see how this works, suppose that the two predictors 'less prenatal care' and 'Clinic B' are denoted by x_1 and x_2, respectively, so that the logistic model takes the form

$$\ln\left(\frac{P}{1-P}\right) = a + b_1 x_1 + b_2 x_2$$

If the predictors are recoded as $x_1 - x_2$ and x_2, the model may be written alternatively as

$$\ln\left(\frac{P}{1-P}\right) = a + b_1(x_1 - x_2) + (b_1 + b_2)x_2$$

In the second formulation the regression coefficient corresponding to the predictor x_2 is the *sum* of the parameters b_1 and b_2, so the odds ratio associated with x_2 is the *product* of the odds ratios. After fitting the logistic model, the odds ratio and confidence interval associated with x_2 will refer to the combination of the two risk factors, as required. The data layout and resulting printout are shown next.

D	Total	$x_1 - x_2$	x_2
3	179	1	0
4	297	0	0
17	214	0	1
2	25	−1	1

```
Logistic Regression Analysis
Response: Died within 1 month
  Name                   Coeff StErr p-value  Odds   95%     CI

  Constant              -4.248 0.442  0.000
  Less PNC - Clin B      0.110 0.561  0.844 1.117 0.372   3.353
  Clinic B               1.809 0.529  0.001 6.107 2.164  17.233

  df:1          Dev:0.043 %(0):96.364 #iter:9  RSq: 0.998
```

TABLE 5.5: Logistic Model giving Combined Effect: Neonatal Mortality

The printout reveals an estimated odds ratio 6.11 for the combined effect of less PNC and Clinic B, with confidence interval (2.16, 17.23).

The above method may be extended to handle effect modification. In this case the logistic model contains an additional interaction term, and takes the form

$$\ln\left(\frac{P}{1-P}\right) = a + b_1 x_1 + b_2 x_2 + b_3 x_1 x_2$$

The odds ratio for the combined effect of the two risk factors is obtained, as before, by comparing the combination ($x_1 = 1$, $x_2 = 1$) with ($x_1 = 0$, $x_2 = 0$), and is thus given by the expression $\exp(b_1 + b_2 + b_3)$. If the predictors are recoded as $x_1 - x_2$, $x_2 - x_1 x_2$, and $x_1 x_2$, an alternative way of writing the model is

$$\ln\left(\frac{P}{1-P}\right) = a + b_1(x_1 - x_2) + (b_1 + b_2)(x_2 - x_1 x_2) + (b_1 + b_2 + b_3)x_1 x_2$$

Consequently you may get a confidence interval for the odds ratio corresponding to the combined effect of the two risk factors by fitting the logistic model to the recoded data and focusing on the term involving $x_1 x_2$ in printout.

Example 4 provides an illustration of the method. The data layout and printout are as follows.

D	Total	$E - F$	$F - E \times F$	$E \times F$
155	335	0	0	1
20	100	−1	1	0
45	65	1	0	0
180	300	0	0	0

```
Logistic Regression Analysis
Response: D
  Col Name        Coeff   StErr p-value  Odds   95%     CI
```

Col	Name	Coeff	StErr	p-value	Odds	95%	CI
0	Constant	0.405	0.118	0.001			
3	E-F	0.406	0.293	0.167	1.500	0.844	2.667
4	F-E*F	-1.386	0.436	0.001	0.250	0.106	0.588
5	E*F	-0.555	0.161	0.001	0.574	0.419	0.787

```
df:0            Dev:0   %(0):50   #iter:8      RSq: 1
```

The estimated odds ratio for the combined effect is thus 0.57 with the 95% confidence interval (0.42, 0.79).

The methods considered in this section extend in a straightforward way to situations involving further stratification variables. In the next section the Berkeley graduate admissions data are analysed using logistic regression.

3: Modelling Stratified Data

The discussion given in the preceding section relates to a dichotomous exposure and a dichotomous stratification variable which alternatively could be viewed as a further risk factor. Consider now a more general situation in which the stratification variable has several categories, such as the data cited at the end of Section 4 of Chapter 4, relating acceptance into a graduate department at Berkeley to the gender of the applicant. The data layout is as follows:

Accepted	Total	Gender	Dept
512	825	1	A
89	108	0	A
353	560	1	B
17	25	0	B
120	325	1	C
202	593	0	C
138	417	1	D
131	375	0	D
53	191	1	E
94	393	0	E
22	373	1	F
24	341	0	F

Note that the *gender* variable is coded as 1 for men and 0 for women. Since all of the predictor variables in the model given by Equation (5.6) are strictly numerical, the department of admission must also be recoded as a set of indicator variables before fitting the model, giving the expanded data layout shown in Table 5.6 below.

Accepted	Total	Gender (men)	Department (A B C D E F)					
512	825	1	1	0	0	0	0	0
89	108	0	1	0	0	0	0	0
353	560	1	0	1	0	0	0	0
17	25	0	0	1	0	0	0	0
120	325	1	0	0	1	0	0	0
202	593	0	0	0	1	0	0	0
138	417	1	0	0	0	1	0	0
131	375	0	0	0	0	1	0	0
53	191	1	0	0	0	0	1	0
94	393	0	0	0	0	0	1	0
22	373	1	0	0	0	0	0	1
24	341	0	0	0	0	0	0	1

TABLE 5.6: Graduate Admissions Data in Logistic Regression Format

Thus there are nine columns in the expanded data array, the last seven corresponding to predictor variables. Table 5.7 shows printout obtained after fitting the model given by Equation (5.6) to these data.

All the variables are included in the model with the exception of the indicator variable for Department A. (You must omit one of the indicator variables corresponding to the department of admission, because only five of the six are needed to define the department uniquely.)

```
Logistic Regression Analysis
Response: Accepted
  Col Name      Coeff  StErr p-value  Odds   95%    CI
```

Col	Name	Coeff	StErr	p-value	Odds	95%	CI
0	Constant	0.682	0.099	0.000			
3	Men	-0.100	0.081	0.217	0.905	0.772	1.060
5	B	-0.043	0.110	0.693	0.958	0.772	1.188
6	C	-1.263	0.107	0.000	0.283	0.230	0.349
7	D	-1.295	0.106	0.000	0.274	0.223	0.337
8	E	-1.739	0.126	0.000	0.176	0.137	0.225
9	F	-3.306	0.170	0.000	0.037	0.026	0.051

```
df:5    Dev:20.204  %(0):61.224 #iter:8  RSq: 0.977
```

TABLE 5.7: Printout from Logistic Model: Berkeley Graduate Admissions

There are 5 degrees of freedom because, with six strata corresponding to the six departments and two proportions (for the men and the women) for each department, you would need 12 parameters to fit the data exactly. However, only seven parameters are used in the model shown in Table 5.7, these being the constant (a), the coefficient for gender (b_1), and the coefficients for the five departments B–F.

The numbers in the second row of the printout table (labelled Men) are those of most interest, since they relate to the exposure variable. The estimated odds ratio is 0.905, with a 95% confidence interval (0.772, 1.060). This is the common odds ratio for gender after adjusting for the stratification variable (department), and may be compared with the estimate of 0.928 obtained in Section 4 of the preceding chapter.

The numbers in the other rows of the printout table relate to the odds of being accepted into the various departments compared to Department A (the omitted category) after adjusting for gender. You can see that the odds of getting into Departments C–F are relatively small compared to Departments A and B.

Recall the graph in Figure 4.2, which shows that the ratio of the odds of acceptance for men versus women in Department A differs from the odds ratios for the other departments, a goodness-of-fit test confirming this result $(p = 0.003)$. Recall that the printout from the logistic regression model also contains a goodness-of-fit statistic: *provided the counts in the strata are reasonably large*, the deviance given in the bottom line of a logistic regression printout may be compared with a chi-squared distribution with degrees of freedom given by df. The value obtained for the deviance in the present case is 20.204 with 5 degrees of freedom, giving the p-value 0.001.

The fit of the common odds ratio model is poor, and it is instructive to find out how to improve the situation. The model may be used to obtain *predicted* probabilities for each *cell*, that is, for each combination of predictor variables. These probabilities are obtained simply by substituting the estimates of the parameters into the right-hand side of Equation (5.6), and may be compared with the observed proportions. Furthermore a z-statistic may be computed for each cell proportion by dividing the residual proportion by an estimate of its standard error using the formula

$$SE(p) = \sqrt{\frac{\hat{p}(1-\hat{p})}{n}} \qquad (5.8)$$

for the standard error of a binomial proportion. (This formula is comparable to Equation (2.1), but uses the predicted proportion $\hat{p}$ instead of the observed proportion p.) The resulting z-statistics are given below in Table 5.8.

p	$\hat{p}$	z	Gender	Dept.
0.621	0.642	−1.25	1	A
0.824	0.664	3.52	0	A
0.63	0.631	−0.06	1	B
0.68	0.654	0.27	0	B
0.369	0.336	1.26	1	C
0.341	0.359	−0.92	0	C
0.331	0.329	0.08	1	D
0.349	0.351	−0.09	0	D
0.277	0.239	1.24	1	E
0.239	0.258	−0.84	0	E
0.059	0.062	−0.21	1	F
0.07	0.068	0.21	0	F

TABLE 5.8: Predicted Proportions and z-scores: Graduate Admissions

The z-statistics obtained in this way are similar to the components of the chi-squared statistic for comparing observed and expected counts in a contingency table. In fact squaring and adding these z-statistics gives an overall measure of goodness-of-fit for the logistic model, which may be compared with a chi-squared distribution in which the number of degrees of freedom is, as usual, equal to the number of cells minus the number of parameters in the model. In the present case this sum of squares turns out to be 18.82, giving a p-value of 0.002 when compared with the chi-squared distribution with 5 degrees of freedom. This is quite similar to the goodness-of-fit result based on the deviance.

Even though each z-statistic obtained using the method just described gives a measure of how well the model predicts the proportion of outcomes in a particular cell, it is important to realise that these z-statistics are not mutually independent. (If they were, squaring and adding them would give a chi-squared statistic with 12 degrees of freedom rather than 5.) Even so, the contributions for individual strata are obtained by summing the squares within these strata. Thus, for example, the contribution to the squared error from Department A is $(-1.25)^2 + (3.52)^2 = 13.95$, while the contribution from Department B is $(-0.06)^2 + (0.27)^2 = 0.08$.

The contributions to the overall squared error from the six departments, after fitting the logistic regression model listed in Table 5.7, are plotted (on a square root scale to compensate for squaring the errors) in Figure 5.2. This graph, like its odds ratio counterpart Figure 4.2, reveals that the contribution to the error from Department A is unusually large.

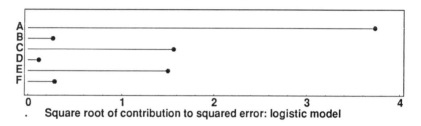

Square root of contribution to squared error: logistic model

FIGURE 5.2: Error Components for Logistic Model: Graduate Admissions

Given the poor fit of the common odds ratio model, you could look for a model with sufficiently many additional parameters to give an acceptable fit. Since Department A is largely responsible for the poor fit, it is reasonable to fit a model in which the odds ratio for this department is allowed to differ from the others. This can be done by including an additional predictor variable for the interaction between gender and Department A. This predictor is simply the product of the corresponding indicator variables. The printout after fitting this more complex model is shown in Table 5.9.

Comparing Tables 5.7 and 5.9, you can see that the deviance has decreased to 2.56, and comparing this with the chi-squared distribution with 4 degrees of freedom, the p-value is 0.63, indicating a good fit.

The odds ratios are presented in Table 5.10, where the referent group is taken to be women applying to Department A. Thus the numbers in the first row are the odds ratios of acceptance for a woman to each department compared to Department A, as shown in the printout in Table 5.9. The numbers in the second row are the odds

ratios of acceptance to each department for a man compared to a
woman applying to Department A. For Departments B–F, these odds
ratios are obtained by multiplying the odds ratios in the first row by
1.031 (the common odds ratio of acceptance for a man compared to a
woman). For Department A, a further factor of 0.339 is needed (the
odds ratio associated with the interaction term for male gender and
Department A).

```
Logistic Regression Analysis
Response: Accepted
 Col Name       Coeff    StErr p-value  Odds    95%    CI

   0 Constant   1.544    0.253  0.000
   3 Men        0.031    0.087  0.724  1.031 0.870 1.222
  10 Men*A     -1.083    0.277  0.000  0.339 0.197 0.582
   5 B         -1.031    0.279  0.000  0.357 0.206 0.617
   6 C         -2.171    0.264  0.000  0.114 0.068 0.191
   7 D         -2.225    0.268  0.000  0.108 0.064 0.183
   8 E         -2.644    0.272  0.000  0.071 0.042 0.121
   9 F         -4.236    0.299  0.000  0.014 0.008 0.026

 df:4      Dev:2.556  %(0):61.224  #iter:9  RSq: 0.997
```

TABLE 5.9: Printout with Interaction Term included in Model

	Department					
	A	B	C	D	E	F
women	(1)	0.357	0.114	0.108	0.071	0.014
men	0.350	0.368	0.118	0.111	0.073	0.014
	0.350	1.031	1.031	1.031	1.031	1.031
	(0.349	0.803	1.133	0.921	1.222	0.828)

TABLE 5.10: Odds Ratios based on Logistic Model: Graduate Admissions

The numbers in the bottom line are the odds ratios of acceptance for a man compared to a woman applying to the same department. These odds ratios are the same except for Department A. They
may be compared with the odds ratios computed directly from the
data, shown in parentheses. (Refer to Table 4.7.) The conclusion from
the earlier analysis was that Departments B–F have a common odds
ratio for the effect of male to female gender with a 95% confidence
interval (0.87, 1.22). This interval agrees with the confidence interval
for the effect of gender shown in Table 5.9.

It is instructive to note that separate odds ratios for each
stratum may be obtained from a logistic model by replacing the common predictor for gender by separate predictors for the effect of

gender within each department. These predictors correspond to inter-actions between male gender and the indicators for each department. The printout shown in Table 5.11 is thus obtained.

```
Logistic Regression Analysis
Response: Accepted
 Col Name      Coeff    StErr  p-value  Odds   95%    CI

   0 Constant  1.542    0.253  0.000
   5 B        -0.788    0.498  0.113  0.455 0.171 1.206
   6 C        -2.203    0.267  0.000  0.111 0.065 0.186
   7 D        -2.164    0.275  0.000  0.115 0.067 0.197
   8 E        -2.699    0.279  0.000  0.067 0.039 0.116
   9 F        -4.123    0.330  0.000  0.016 0.008 0.031
  10 Men*A    -1.050    0.263  0.000  0.350 0.209 0.585
  11 Men*B    -0.220    0.438  0.615  0.802 0.340 1.891
  12 Men*C     0.125    0.144  0.386  1.133 0.855 1.502
  13 Men*D    -0.082    0.150  0.585  0.921 0.686 1.237
  14 Men*E     0.200    0.200  0.317  1.222 0.825 1.809
  15 Men*F    -0.189    0.305  0.536  0.828 0.455 1.506

 df:0        Dev:0   %(0):61.224   #iter:10      RSq: 1
```

TABLE 5.11: Saturated Logistic Model: Berkeley Graduate Admissions

This model is a *saturated* model, because it has no degrees of freedom. The odds ratios for the interaction terms are the same as those computed directly from the data (Table 4.7). However, the model with 4 degrees of freedom (Table 5.9) provides a perfectly acceptable fit.

In this section you have seen how a logistic regression model may be used to handle a stratification variable which could be an effect modifier. In the next section you will see how the model is applied to situations involving multiple risk factors.

4: Multiple Risk Factors

The data layout for fitting the logistic regression model to the oes-ophageal cancer case-control study (Section 6 of Chapter 4) comprises 54 rows (or *cells*) as shown in Table 5.12.

This data layout is similar to that for the example discussed in the preceding section, so that y and n refer to the number of disease outcomes and the total in each strata, while R_1, R_2 and R_3 refer to the risk factors *tobacco consumption* and *alcohol consumption* and the stratification variable *age group*, respectively. These variables are coded as follows:

Tobacco: 0 = 0–9 g/day, 1 = 10–19 g/day, 2 = 20+ g/day;
Alcohol: 0 = 0–39 g/day, 1 = 40–79 g/day, 2 = 80+ g/day;
Age Group: 0 = 25–34 years, 1 = 35–44, 2 = 45-54, 3 = 55–64,
           4 = 65–74, 5 = 75+ years.

y	n	R_1	R_2	R_3	y	n	R_1	R_2	R_3
0	40	0	0	0	2	49	0	0	3
0	27	0	1	0	9	40	0	1	3
0	3	0	2	0	14	28	0	2	3
0	10	1	0	0	3	22	1	0	3
0	7	1	1	0	6	21	1	1	3
1	2	1	2	0	14	22	1	2	3
0	11	2	0	0	7	18	2	0	3
0	11	2	1	0	7	23	2	1	3
0	5	2	2	0	14	19	2	2	3
0	60	0	0	1	5	48	0	0	4
0	35	0	1	1	17	34	0	1	4
2	14	0	2	1	9	17	0	2	4
1	14	1	0	1	4	14	1	0	4
3	23	1	1	1	3	10	1	1	4
0	9	1	2	1	5	14	1	2	4
0	15	2	0	1	2	9	2	0	4
1	22	2	1	1	5	9	2	1	4
2	7	2	2	1	5	6	2	2	4
1	46	0	0	2	1	18	0	0	5
6	38	0	1	2	2	5	0	1	5
7	20	0	2	2	3	3	0	2	5
0	18	1	0	2	2	6	1	0	5
4	21	1	1	2	1	3	1	1	5
9	18	1	2	2	2	2	1	2	5
0	14	2	0	2	1	3	2	0	5
10	22	2	1	2	1	4	2	1	5
9	16	2	2	2	0	0	2	2	5

TABLE 5.12: Data Layout for Logistic Model: Oesophageal Cancer Study

Since the risk factors are ordinal variables, they should be replaced by indicator variables before fitting the logistic regression model, giving a data array with 14 columns in all, and it is then necessary to omit one indicator variable from each of the three sets. The printout in Table 5.12 is obtained as a result of fitting the model after omitting the indicator variables corresponding to the lowest consumption levels of tobacco and alcohol and the age group 25–34 years.

The printout in Table 5.13 may be compared with that in Table 5.9. Instead of having a single coefficient encapsulating the effect of the exposure variable of interest, there are now four coefficients, two for tobacco consumption and another two for alcohol consumption. The odds ratios corresponding to these coefficients may be compared with those based on the Mantel–Haenszel formula given in the margins of Table 4.15. Thus the odds ratios for the two levels of tobacco consumption, from the model, are 1.48 and 2.57, respectively, compared with Mantel–Haenszel estimates of 1.51 and 2.73 based on Equation (4.6). Similarly, the printout gives model-based estimates for the odds ratios corresponding to the two levels of alcohol consumption as 3.95 and 11.57, compared with Mantel–Haenszel estimates of 3.76 and 11.06, respectively.

```
Logistic Regression Analysis
Response: oesophageal cancer
Col Name          Coeff  StErr p-value    Odds   95%      CI
```

Col	Name	Coeff	StErr	p-value	Odds	95%	CI
0	Constant	-6.304	1.038	0.000			
4	Tob 10-19	0.391	0.224	0.081	1.478	0.953	2.293
5	Tob 20+	0.944	0.228	0.000	2.570	1.645	4.015
7	Alc 40-79	1.373	0.247	0.000	3.946	2.432	6.403
8	Alc 80+	2.448	0.259	0.000	11.566	6.967	19.202
10	Age 35-44	1.564	1.071	0.144	4.777	0.586	38.947
11	Age 45-54	3.223	1.028	0.002	25.110	3.346	188.46
12	Age 55-64	3.798	1.024	0.000	44.624	5.994	332.21
13	Age 65-74	4.242	1.032	0.000	69.532	9.207	525.09
14	Age 75+	4.368	1.080	0.000	78.906	9.494	655.77

```
df:43    Dev:55.958   %(0):79.487   #iter:12    RSq: 0.821
```

TABLE 5.13: Printout from Multiple Risk Factor Model

In this model age is a risk factor and a possible confounder. If it is omitted, you get the printout shown in Table 5.14.

```
Logistic Regression Analysis
Response: oesophageal cancer
Col Name          Coeff  StErr p-value    Odds   95%      CI
```

Col	Name	Coeff	StErr	p-value	Odds	95%	CI
0	Constant	-2.853	0.213	0.000			
4	Tob 10-19	0.364	0.210	0.083	1.439	0.954	2.170
5	Tob 20+	0.709	0.208	0.001	2.032	1.352	3.053
7	Alc 40-79	1.204	0.234	0.000	3.333	2.107	5.272
8	Alc 80+	2.384	0.240	0.000	10.849	6.773	17.380

```
df:48    Dev:170.467   %(0):79.487   #iter:9    RSq:0.454
```

TABLE 5.14: Logistic Model for Oesophageal Cancer with Age omitted

Comparing the printouts in Tables 5.13 and 5.14, the main difference is in the bottom line, where the deviance increases by more than a factor of three when age is omitted. In this case failure to stratify by age gives a poorer fit, but does not substantially change the estimates of the odds ratios for the risk factors of interest. In other words, age is a risk factor for the disease, but not a confounder for the effects of alcohol and tobacco.

The logistic model may be used to calculate an odds ratio for any two exposure groups, and Equation (5.7) gives the formula. Suppose, for example, based on the printout shown in Table 5.13, you wish to compare the odds of disease for subjects consuming 10–19 grams/day of tobacco *and* 40–79 grams/day of alcohol with those in the baseline group (0–9 grams of tobacco and 0–39 grams of alcohol). Since the age groups are presumed to be the same for the two exposure groups, the only predictors contributing to the formula (5.7) are x_4 and x_7, so the formula gives $\exp(b_4 + b_7)$, that is, $\exp(0.391 + 1.373)$, which equals 5.83. (This result may also be obtained by multiplying together the odds ratios 1.478 and 3.946.)

The effect of two risk factors is thus obtained by *adding* their contributions to the linear model, which corresponds to *multiplying* their odds ratios, since the odds ratio is obtained by exponentiating the coefficient in the linear model as in Equation (5.5). The odds ratios for the various exposure groups (compared with the referent exposure group having the lowest tobacco and alcohol consumption) may thus be estimated using the model, and are given in Table 5.15.

Alcohol Consumption	Tobacco Consumption 0-9 g/day	10-19 g/day	20+ g/day	
0-39 g/day	(1.0)	1.48	2.57	(1.0)
40-79 g/day	3.95	5.83	10.14	3.95
80+ g/day	11.57	17.09	29.72	11.57
	(1.0)	1.48	2.57	

TABLE 5.15: Odds Ratios from Logistic Model with No Interactions

If the joint effect of two risk factors on the odds ratio is not multiplicative, a more complex model is needed. This model will have a component for each combination of levels of the two risk factors. When the data array is reformulated in this way (so that the two sets of three indicator variables are replaced by one set of nine *interaction* terms), the printout in Table 5.16 is obtained. The interaction terms are labelled so that T0*A0 refers to 0–9 grams/day of tobacco and 0–39 grams/day of alcohol, T0*A1 to 0-9 grams/day of tobacco and 40-79 grams/day of alcohol, etc. Note that the interaction term corresponding to the lowest level of consumption of both tobacco and alcohol has been omitted, so that the odds ratios again refer to this baseline level.

```
Logistic Regression Analysis
Response: oesophageal cancer
 Col Name        Coeff  StErr p-value   Odds    95%        CI
─────────────────────────────────────────────────────────────────
   0 Constant   -6.801  1.072  0.000
  10 T1*A0        1.290  0.490  0.008    3.632   1.390    9.486
  11 T2*A0        1.703  0.500  0.001    5.491   2.062   14.617
  12 T0*A1        2.021  0.402  0.000    7.545   3.429   16.599
  13 T1*A1        2.159  0.453  0.000    8.664   3.563   21.068
  14 T2*A1        2.608  0.437  0.000   13.578   5.768   31.962
  15 T0*A2        2.993  0.423  0.000   19.949   8.705   45.714
  16 T1*A2        3.183  0.439  0.000   24.109  10.199   56.986
  17 T2*A2        3.959  0.478  0.000   52.406  20.543  133.689
   4 Age 35-44    1.596  1.072  0.137    4.932   0.603   40.339
   5 Age 45-54    3.241  1.030  0.002   25.568   3.399  192.351
   6 Age 55-64    3.806  1.026  0.000   44.967   6.022  335.748
   7 Age 65-74    4.256  1.033  0.000   70.497   9.309  533.901
   8 Age 75+      4.422  1.084  0.000   83.255   9.955  696.311
─────────────────────────────────────────────────────────────────
 df:39     Dev:50.128    %(0):79.487    #iter:12    RSq:0.839
```

TABLE 5.16: Model for Oesophageal Cancer with Interactions Included

The printout in Table 5.16 may be interpreted in much the same way as that shown in Table 5.13. There are now eight estimated odds ratios, which may be compared with the Mantel–Haenszel estimates given in the body of Table 4.15. For example the logistic model gives 8.664 for the odds ratio of T1*A1 compared with T0*A0, whereas the Mantel–Haenszel estimate is 8.63. These odds ratios are given in Table 5.17.

Alcohol	Tobacco Consumption		
Consumption	0-9 g/day	10-19 g/day	20+ g/day
0-39 g/day	(1.0)	3.63	5.49
40-79 g/day	7.54	8.66	13.58
80+ g/day	19.95	24.11	52.41

TABLE 5.17: Odds Ratios from Logistic Model including Interactions

Comparing the two tables of estimated odds ratios based on the alternative models, you can see that those based on the more extensive model are generally larger. The question arises: 'which model should be preferred?'

The values of the deviances given in the bottom line of a printout may be used to compare two logistic regression models. Provided the two models are *hierarchical* (that is, one model is a subset of the other) a *p*-value is found by comparing the difference between the deviances with a chi-squared distribution having degrees of freedom equal to the number of additional parameters in the larger model. A small p-value indicates that the larger model provides a better fit, whereas a relatively large p-value indicates a preference for the more parsimonious model.

In the present case deviance for the model the difference in deviances is $55.96 - 50.13$, or 5.83, and comparing this to a chi-squared distribution with 4 degrees of freedom, the p-value is 0.21. Since this p-value is not small (by the conventional 0.05 criterion), the model containing the interaction terms does not provide strong evidence for a superior fit.

Although the deviance difference may be used to compare two models with each other, you still need to determine if the fit of the preferred model is satisfactory. Provided there are not too many strata containing small counts, the deviance itself can provide an indication of the goodness-of-fit of the model, simply by comparing it with a chi-squared distribution with df degrees of freedom. Using this criterion the simpler model given in Table 5.13 has a p-value 0.089 (based on a deviance of 55.96 with 43 degrees of freedom) while the model with the interaction terms given in Table 5.16 has a p-value 0.109 (based on a deviance of 50.13 with 39 degrees of freedom), so in each case the fit is not particularly good, though acceptable given that the p-values exceed 0.05. By the same token the fit of the model given in Table 4.10 in which the age effects are omitted is very poor.

It is also worth examining the z-statistics (often called *standardised residuals*) which are obtained by dividing the residual proportions by their theoretical standard error estimates (Equation (5.8)). Ill-fitting strata are highlighted when these standardised residuals are plotted against normal scores. Since the residuals in this plot are supposed to be standardised to have unit standard deviation, it is useful to compare them with a straight line having unit slope. Figure 5.3 shows such a plot for the simpler model without the interaction terms. Since strata containing larger numbers of subjects carry more weight, the areas of the circles used as symbols in this plot are in direct proportion to the sample sizes of these cells.

Two outliers are evident in Figure 5.3, one positive and one negative. The positive extreme corresponds to the stratum in Table 4.14 defined by age group 25–34, tobacco consumption 10–19 grams per day, and alcohol consumption 80+ grams/day, where just one case and one control were observed. For this cell the model predicts the proportion of cases to be 0.03, or 0.06 out of 2, whereas the observed proportion was 1 out of 2. This discrepancy, when standardised, gives a z-statistic close to 4, but does not in itself provide evidence of a poor fit since the cell size is so small. The negative extreme occurs in the age group 65–74, tobacco consumption level 10–19 grams/day and alcohol consumption level 80+ grams/day, where only 5 cases were observed with 14 controls but 9.59 cases and 4.41 controls are expected on the basis of the model, giving a z-statistic −2.64. Given the large number of cells it is not surprising that one of them would give a result as extreme as this, so again the result is not particularly remarkable.

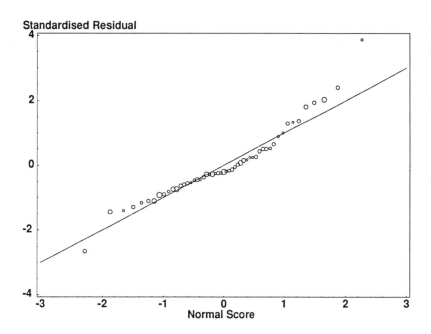

FIGURE 5.3: Plot of Standardised Residuals vs Normal Scores

Table 5.18 gives an alternative formulation of the larger model. The difference between this model and that given in Table 5.16 is that the 8 terms encapsulating the effects of the tobacco and alcohol risk factors are now coded as 4 main effects (as in the smaller model reported in Table 5.13) plus 4 further interactions, rather than as pure combinations of the two risk factors.

The smaller model contains separate effects for the two risk factors. As noted earlier, this model assumes that the joint risk associated with additional consumption of both tobacco and alcohol is a multiplicative combination of the separate effects. Compared with the more extensive model involving individual parameters for each combination of tobacco and alcohol consumption, it gives smaller odds ratios for the various exposure groups compared with the baseline exposure group.

```
Logistic Regression Analysis
Response: oesophageal cancer
 Col Name        Coeff  StErr p-value   Odds    95%      CI
```

Col	Name	Coeff	StErr	p-value	Odds	95%	CI
0	Constant	-6.801	1.072	0.000			
4	Tob 10-19	1.290	0.490	0.008	3.632	1.390	9.486
5	Tob 20+	1.703	0.500	0.001	5.491	2.062	14.617
7	Alc 40-79	2.021	0.402	0.000	7.545	3.429	16.599
8	Alc 80+	2.993	0.423	0.000	19.949	8.705	45.714
9	T1.A1	-1.151	0.604	0.057	0.316	0.097	1.033
10	T1.A2	-1.100	0.605	0.069	0.333	0.102	1.089
11	T2.A1	-1.115	0.598	0.062	0.328	0.101	1.059
12	T2.A2	-0.737	0.636	0.247	0.478	0.138	1.665
14	Age 35-44	1.596	1.072	0.137	4.932	0.603	40.340
15	Age 45-54	3.241	1.030	0.002	25.568	3.399	192.35
16	Age 55-64	3.806	1.026	0.000	44.967	6.022	335.75
17	Age 65-74	4.256	1.033	0.000	70.497	9.309	533.91
18	Age 75+	4.422	1.084	0.000	83.256	9.955	686.32

```
 df:39     Dev:50.128   %(0):79.487  #iter:12   RSq:0.839
```

TABLE 5.18: Alternative Model with Interactions

The difference between the two models is perhaps best understood by comparing the printouts in Tables 5.16 and 5.18. If the multiplicative model were correct, the coefficients of the interaction terms (labelled T1*A1, T1*A2, T2*A1 and T2*A2) in Table 5.18 would all be close to 0, and the odds ratios for the corresponding exposure groups would be obtained by multiplying the odds ratios corresponding to the separate components (or by adding and exponentiating coefficients of the main effects listed above the interaction terms).

The coefficients of the terms T1*A1, T1*A2, T2*A1 and T2*A2 in Table 5.18 thus reflect the departure from the simple multiplicative (or independent effects) model. A negative coefficient signifies that the joint effect of the two risk factors, at the given levels, is *less* that what this model would predict, and conversely, a positive coefficient indicates that the joint effect is greater than that obtained by simply multiplying the odds. Since all of these relevant coefficients in Table 5.18 are negative and all but one have small *p*-values, the former conclusion may be reasonable. Based on the larger model, the joint effect of increased consumption of tobacco and alcohol on the disease in this study is less than what would occur if the risks were acting independently.

Of course the above conclusion should be tempered by the fact that the larger model does not provide a better fit than the smaller model.

Summary

In this chapter a basic statistical model for dichotomous outcomes – the logistic regression model – has been introduced. This model gives a formula which expresses the risk of an event in terms of a linear function of predictor variables, and thus is similar in many respects to the multiple regression model given by Equation (3.6).

The logistic model gives a particularly simple representation for the logarithm of the odds ratio associated with a risk factor, and when fitted to data involving dichotomous outcomes and exposures, it automatically provides estimates of odds ratios and confidence intervals for specific combinations of risk factors. Thus logistic regression analysis provides an alternative approach to the Mantel–Haenszel methods for analysing 2-by-2 contingency tables covered in Chapter 4.

Confounding Examples

The logistic regression model was used to reanalyse the four confounding examples introduced in Section 2 of Chapter 4, and you saw how effect modification could be accommodated as an interaction term expressible as the product of two predictors in the model. This approach also gives a confidence interval for a combination of risk factors.

Modelling Stratified Data

As an illustration of modelling stratified data, various logistic models were fitted to the 1973 Berkeley graduate admissions data (introduced in Section 3 of Chapter 4). The results agree with those obtained from the Mantel–Haenszel analysis, namely that the sex discrimination favoured men rather than women and was confined to Department A.

Multiple Risk Factors

A case-control study of alcohol and tobacco consumption as risk factors for oesophageal cancer, and analysed extensively by Breslow and Day (1980), provided a illustration of logistic modelling of multiple risk factors, including the assessment of the need to include interaction terms to account for possible departures from independence of effects.

Further aspects of logistic regression are covered in the next chapter. These include the analysis of risk factors which vary continuously from subject to subject and thus require a different data layout to the one assumed in the present chapter, and applications of Poisson regression, which arises as the special case of logistic regression when the risk of an event is small.

The books by Cox and Snell (1989), Hosmer and Lemeshow (1989), Dobson (1990), Collett (1991), and Kleinbaum (1994), provide extensive discussion of the theory and application of logistic regression.

Exercises

Exercise 5.1: Use logistic regression to analyse the physical activity/ myocardial infarct case-control study data cited in Exercise 4.4 from Chapter 4.

Exercise 5.2: Analyse the data from Bartlett's plum root cuttings (Exercise 4.6 from Chapter 4) using logistic regression. Is there evidence that the effects associated with cutting length and planting time are not independent?

Exercise 5.3: The following data, reported by Maag and Behrens (1989), are taken from a cross-sectional study of depression (classified as 'high' or 'low') in two groups of adolescents, classified as seriously emotionally disturbed (SED) or learning disabled (LD), and stratified by age and sex. Analyse these data using logistic regression.

Age	Group	Sex	Depression Low	Depression High
12-14	LD	male	79	18
		female	34	14
	SED	male	14	5
		female	5	8
15-16	LD	male	63	10
		female	26	11
	SED	male	32	3
		female	15	7
17-18	LD	male	36	13
		female	16	1
	SED	male	36	5
		female	12	2

Exercise 5.4: After Australia had convincingly defeated India in the first two cricket test matches at Brisbane and Melbourne in December 1991, Indian former captain Sunil Gavaskar claimed in a newspaper column that Australian umpires were biased in favour of the home team, particularly with regard to controversial 'leg before wicket'

(LBW) decisions. On January 5, 1992, the Sydney *Sun-Herald* news-paper (page 43) noted that since 1977, Australian umpires had upheld only 95 appeals for LBW from 855 Test Match innings (11.11%) against Australian top-order batsmen, compared with 121 from 874 innings (13.84%) against opponent top-order batsmen. Gavaskar's response was 'the figures speak for themselves'.

Settle the issue by fitting a logistic regression model.

Exercise 5.5 (continuation of 5.4): The *Sun-Herald* article gave a more complete data listing of the LBW decisions as follows.

Test Series	Australia	Opponent
vs India 1977-78	7/60	3/54
vs England 1978-79	5/72	18/68
vs Pakistan 1978-79	2/22	1/24
vs England 1979-80	3/34	6/36
vs W.Indies 1980-81	6/36	3/31
vs N.Zealand 1980-81	2/30	5/36
vs India 1980-81	1/30	6/36
vs Pakistan 1981-82	3/32	5/30
vs W.Indies 1981-82	4/36	4/36
vs England 1982-83	6/56	3/60
vs Pakistan 1983/84	6/38	5/53
vs W.Indies 1984-85	6/54	5/52
vs N.Zealand 1985-86	4/36	3/30
vs India 1985-86	3/32	1/22
vs England 1986-87	7/59	8/51
vs N.Zealand 1987-88	6/27	4/36
vs England 1987-88	0/10	0/6
vs Sri Lanka 1987-88	1/6	4/12
vs W.Indies 1988-89	6/59	8/57
vs N.Zealand 1989-90	1/6	3/12
vs Sri Lanka 1989-90	3/24	4/18
vs Pakistan 1989-90	6/28	7/30
vs England 1990-91	6/50	13/60
vs India 1991-92	1/18	2/24

Use these data to further investigate the issue of umpire bias.

Exercise 5.6: Weinberg and Gladen (1986) considered the following data, obtained from women who succeeded in getting pregnant and who were asked how many cycles it took them. The women were also classified as smokers if they averaged at least one cigarette per day during at least the first cycle of their pregnancy attempt.

Use logistic regression to model the risk of getting pregnant as a function of the number of cycles and smoking habit, for women in this target population.

Cycles	Smokers	Non-smokers
1	29	198
2	16	107
3	17	55
4	4	38
5	3	18
6+	31	70

Exercise 5.7: Fienberg (1977) quotes a data set from Reiss et al (1975) on pre-marital contraceptive usage. These data, listed in the following table, come from undergraduate women university students one group of whom had attended the university contraceptive clinic and one group of whom had not.

Are virginity and attitude related to clinic use? Is there reason to believe that virginity and attitude are themselves related? Answer these questions using logistic regression analysis.

Attitude on extra-marital coitus	Use of clinic			
	Yes		No	
	Virgin	Non-virgin	Virgin	Non-virgin
Always wrong	23	127	23	18
Not always wrong	29	112	67	15

Exercise 5.8: Use logistic regression to analyse the data from the case-control study of risk factors for myocardial infarction cited in Exercise 4.5.

References

Breslow, N.E. & N.E. Day (1980): *Statistical Methods in Cancer Research: Volume I − The Analysis of Case-Control Studies.* IARC. Lyon.

Collett, D. (1991): *Modelling Binary Data*. Chapman & Hall. London.

Cox, D.R. & E.J Snell (1989): *Analysis of Binary Data* (2nd ed.). Chapman & Hall. London.

Dobson, A.J. (1990): *An Introduction to Generalized Linear Models* (2nd ed.). Chapman & Hall. London.

Fienberg, S.E. (1977): *The Analysis of Cross-Classified Categorical Data*. The MIT Press. Cambridge, MA.

Hosmer, D.W. & S. Lemeshow (1989): *Applied Logistic Regression*. Wiley. New York.

Kleinbaum, D.G. (1994): *Logistic Regression: A Self-Learning Text*. Springer-Verlag. Berlin.

Maag, J.W. & J.T. Behrens (1989): Epidemiologic data on seriously emotionally disturbed and learning disabled adolescents: reporting extreme depressive symptomatology, *Behavioral Disorders*, **15**, pages 21–27.

Reiss, I.L., Banwart, A. & Foreman, H. (1975): Premarital contraceptive usage: a study and some theoretical explorations, *Journal of Marriage and the Family*, **37**, pages 619–630.

Weinberg, C.R. & B.C. Gladen (1986): The beta-geometric distribution applied to comparative fecundability studies, *Biometrics*, **42**, pages 547–560.

6
LOGISTIC REGRESSION II

1: Introduction

In Chapters 4 and 5 some basic statistical methods for analysing data from epidemiological study designs were considered. These methods involve the analysis of odds ratios associated with dichotomous outcomes and one or more predictors (including exposure and stratification variables), and logistic regression is the statistical model used for the analysis.

You saw in Chapter 5 that logistic regression is a general method for the analysis of case-control studies, where odds ratios are used to compare risks. The present chapter covers some further important applications of logistic regression. It emphasises methods for the analysis of cross-sectional and cohort studies, where direct estimates of risks, as well as relative risks, are available. The chapter thus provides a detailed account of Poisson regression, a general method used for analysing data from cohort studies (see, for example, Breslow & Day, 1987).

We begin with an alternative way of structuring the data in logistic regression analysis, where each subject in the study has a separate record. This case-by-case data layout is inefficient when dealing with the type of data encountered in Chapter 4, but it is necessary when at least one predictor variable varies continuously.

Next we show how logistic regression may be used to check that subgroups in a study, such as treatment and control groups in a clinical trial, are similar with respect to basic background variables. The same method may be used to check that a sample is representative of a target population.

Poisson regression is introduced as the special case of logistic regression when the outcome risks are small. The method is applied to some illustrative examples including the study of smoking and coronary deaths among British doctors, and the concept of goodness-of-fit of a statistical model is discussed in greater detail. A study of repeated ear infections among swimmers in Sydney provides an interesting application of Poisson regression modelling to case-by-case data. We also show how the Poisson model may be used to analyse data with ordinal outcomes, provided distributional assumptions are met.

Finally we show that Poisson regression provides a method for handling data in which the risk of an event is allowed to vary with time, thus providing an entree to Chapter 7, which is devoted to survival analysis.

2: Case-by-Case Data

In the methods for logistic regression analysis outlined in Chapter 5 the data are assumed to be structured as an array with a separate row corresponding to each combination of predictor variables, as follows:

D	$Total$	X_1	X_2	...	X_p
y_1	n_1	x_{11}	x_{12}	...	x_{1p}
y_2	n_2	x_{21}	x_{22}	...	x_{2p}
.	.	.	.	.	
y_m	n_m	x_{m1}	x_{m2}	...	x_{mp}

In this data layout it is assumed that there are m cells with n_i subjects in cell i of whom y_i experience the outcome event.

In many situations, particularly if the data have arisen from a clinical trial, it is more convenient to have the data stored in a database with a separate record for each subject. This *case-by-case* data structure may also be conceptionally simpler than the grouped layout when some of the predictors have a continuous range of variation.

The case-by-case data layout is just the special case of the grouped data layout which arises when n_i is 1 for each cell. Consider again the simple logistic regression analysis of Bishop's (1969) neonatal mortality data from 715 mothers given in Section 2 of Chapter 5.

		Exposure Prenatal Care (days)	
		0-29	30+
Outcome	Died	20	6
	Survived	373	316

The data layout for fitting the logistic model is as follows:

D	Total	PNC < 30 days
20	393	1
6	322	0

These data were obtained from 715 individuals, and could also be structured with a separate record for each subject as follows. Note that the column indicating the total number of subjects in each cell is redundant and is thus omitted. This data structure is the same as for ordinary multiple regression.

	D	E
	1	1
	1	1
(20 records)	··	··
	1	1
	0	1
	0	1
(373 records)	··	··
	0	1
	1	0
	1	0
(6 records)	··	··
	1	0
	0	0
	0	0
(316 records)	··	··
	0	0

For this example the grouped data structure is a much more efficient method of storing the data. However, if the exposure were measured in exact days, rather than less than or more than one month, the case-by-case structure would be needed. The printout based on fitting the logistic model to the case-by-case data is given in Table 6.1.

```
Logistic Regression Analysis
Response: Neonatal Mortality
  Col Name        Coeff   StErr p-value   Odds   95%    CI

   0 Constant    -3.964   0.412   0.000
   3 Less PNC     1.038   0.472   0.028  2.824 1.120 7.118

df:713      Dev:217.769   %(0):96.364  #iter:9    RSq:0.025
```

TABLE 6.1: Logistic Regression Printout from Case-by-case Data

Compare this printout with that obtained by fitting the model to the grouped data, shown in Table 5.1. The only difference is in the bottom line. The fit of the model to the grouped data is perfect, as it must be when there are only two proportions to be fitted with two parameters. However, the fit of the model to the ungrouped data is less than perfect because here the goodness of fit is defined in terms of the 715 individual proportions, and it is more difficult to predict the outcome for an individual subject than it is to predict the average outcome for a group of individuals.

Since the fit of the model to data grouped by the value of a single dichotomous exposure is always perfect, the goodness-of-fit of the model to such data cannot be assessed. To get a goodness-of-fit measure, the data must be stratified or classified by one or more further risk factors, so that the number of degrees of freedom is greater than 0. In fact the data in the infant mortality study were also classified by clinic as follows:

| | Prenatal care < 30 days | | | | |
	Yes	No		Yes	No
Died within 1 month — Yes	3	4	Yes	17	2
Died within 1 month — No	176	293	No	197	23
	Clinic A			Clinic B	

When the logistic model is fitted to these (grouped) data, the printout shown in Table 5.2 is obtained, showing a deviance of 0.043 and r-squared 0.998. But fitting the same model to the case-by-case data layout gives a deviance of 205.60 and r-squared 0.08. All this means is that the model provides excellent prediction of the proportions within the four exposure groups, but poor prediction of outcomes for individual subjects.

Continuous Predictor Variables

Consider a logistic model containing a single predictor which has a continuous range of variation. From Equation (5.1), the odds of the outcome given that the predictor takes the value $x^{(1)}$ is $\exp(a+b\,x_1)$. It follows that the odds ratio for the value $x^{(1)}$ compared to a referent value $x^{(0)}$ is $\exp(b\,x^{(1)} - b\,x^{(0)})$, and this reduces to just $\exp(b)$ if the difference $x^{(1)} - x^{(0)}$ is 1. This means that the regression coefficient b in the logistic model is the logarithm of the odds ratio associated with two values of the predictor variable that differ by one unit, a straightforward extension of Equation (5.7).

Consider some data from a case-control study of perinatal

mortality undertaken by Dr Hieu in Ho Chi Minh City (see the appendix for a data listing). Take birth weight as a determinant. While birth weight could also be regarded as an outcome in its own right (or as an intervening variable), it is instructive to treat it as a risk factor for the present analysis.

The following printout is obtained by fitting the logistic model to the 412 data records, treating birth weight (in kilograms) as a single continuous predictor. The printout shows an estimated odds ratio of 0.178, which is the ratio of the odds of perinatal death outcome associated with any two birth weights differing by 1 kilogram.

```
Logistic Regression Analysis
Response: Perinatal Mortality
 Col Name          Coeff  StErr  p-value   Odds    95%    CI

   0 Constant      3.443  0.522       0
   3 Birth Wt     -1.726  0.198       0    0.178  0.121  0.263

df:410      Dev:362.807    %(0):75    #iter:10    RSq: 0.217
```

TABLE 6.2: Analysis of Case-by-case Data: A Perinatal Mortality Study

The model assumes that the reduction in the logarithm of the odds of perinatal death associated with a unit increase in birth weight is *linear*. In particular this means that a baby weighing 2000 grams has only 0.178 times the odds of dying as a baby weighing 1000 grams. But it also means that a baby weighing 3000 grams has the same advantage over a baby weighing 2000 grams, and one weighing 4000 grams has the same advantage over one weighing 3000 grams.

It is unlikely that such a simple relation holds. More plausibly, babies within some 'normal' weight range are most likely to survive, and too great a deviation from this range, particularly at the lower end, is likely to increase the risk of perinatal death.

A more realistic model would include a quadratic term in the model, giving the printout in Table 6.3.

```
Logistic Regression Analysis
Response: Perinatal Mortality
 Col Name          Coeff  StErr  p-value   Odds    95%    CI

   0 Constant     11.624  1.711       0
   3 Birth Wt     -8.651  1.268       0    0.000  0.000  0.002
   4 (B.Wt)^2      1.360  0.228       0    3.895  2.489  6.095

df:409      Dev:326.816    %(0):75    #iter:13    RSq: 0.295
```

TABLE 6.3: Inclusion of Quadratic Term: Perinatal Mortality Study

Including the quadratic term improves the fit substantially: the

r-squared has increased from 21.7% to 29.5%, and the deviance has decreased from 362.81 to 326.82. According to this model, the logarithm of the odds of perinatal death is given by the quadratic formula $11.62 - 8.65\,x + 1.36\,x^2$, or $1.36\,(x - 3.18)^2 - 2.12$, where x is the birth weight in kilograms. This expression is minimised when x is 3.18 kilograms, so the model may be interpreted as saying that the logarithm of the odds of the adverse outcome increases in proportion to the square of the amount by which the birth weight differs from 3180 grams.

The model could be made even more flexible by adding further polynomial terms and stopping when there is no worthwhile additional improvement in fit. In the present it turns out that there is no benefit in going beyond the quadratic model, since adding a cubic term reduces the deviance by only 0.22, a statistically insignificant amount.

An alternative method for modelling a continuous predictor variable involves grouping it into categories and using indicator variables to denote membership of the categories. In the present situation it is reasonable to use 500 gram intervals, and taking the group surrounding 3180 grams as the referent group for comparison of odds ratios, the printout shown in Table 6.4 is obtained.

```
Logistic Regression Analysis
Response: Perinatal Mortality
  Col Name            Coeff  StErr p-value  Odds    95%     CI
```

Col	Name	Coeff	StErr	p-value	Odds	95%	CI
0	Constant	-1.930	0.252	0.000			
5	B.Wt:<1500g	4.297	0.654	0.000	73.48	20.38	264.96
6	1500-2000	3.083	0.532	0.000	21.82	7.69	61.87
7	2000-2500	0.918	0.421	0.029	2.50	1.10	5.72
8	2500-3000	-0.048	0.389	0.901	0.95	0.44	2.04
10	3500-4000	-0.198	0.536	0.711	0.82	0.29	2.34
11	4000+	1.370	0.676	0.043	3.94	1.05	14.80

```
 df:405     Dev:333.627    %(0):75     #iter:9    RSq: 0.28
```

TABLE 6.4: Categorical Birth Weight: Perinatal Mortality Study

The model reported in Table 6.4 contains four more parameters than the quadratic model, but it does not fit as well, having an r-squared 28% compared with 29.5% for the quadratic model.

3: Modelling Risks

So far we have focused on logistic regression as a method for analysing odds ratios rather than probabilities (or risks) of events. This approach is used for the analysis of data from case-control studies. However, the logistic model (5.6) gives the risk of an event directly, and in many situations it is the risk itself, rather than a relative risk or odds ratio, that is of primary interest. If the data have arisen from a cross-sectional or cohort study it is possible to estimate the risks associated with specified combinations of determinants.

As an illustration, consider a survey of psychotropic drug-taking behaviour in West London reported by Murray et al (1981). Proportions of drug-takers classified by sex, age group, and General Health Questionnaire (GHQ) result (high or low score) are listed in Table 6.5, together with the printout obtained from fitting a logistic regression model. Note that the deviance is 14.31 with 13 degrees of freedom, so the fit is good ($p = 0.35$).

Sex	Age	GHQ	Drug-takers	Total
m	16–29	low	9	531
m	30–44	low	16	500
m	45–64	low	38	644
m	65–74	low	26	275
m	75+	low	9	90
m	16–29	high	12	171
m	30–44	high	16	125
m	45–64	high	31	121
m	65–74	high	16	56
m	75+	high	10	26
f	16–29	low	12	568
f	30–44	low	42	596
f	45–64	low	96	765
f	65–74	low	52	327
f	75+	low	30	179
f	16–29	high	33	210
f	30–44	high	47	189
f	45–64	high	71	242
f	65–74	high	45	98
f	75+	high	21	60

```
Logistic Regression Analysis
Response: Drug-taker
 Col Name              Coeff  StErr p-value  Odds   95%    CI

   0 Constant         -4.005  0.151          0
   1 Female            0.628  0.096          0 1.873 1.553 2.259
   3 GHQ high          1.414  0.090          0 4.111 3.443 4.908
   7 Age 30-44         0.768  0.161          0 2.154 1.571 2.954
   8 Age 45-64         1.311  0.148          0 3.710 2.777 4.955
   9 Age 65-74         1.736  0.162          0 5.673 4.128 7.797
  10 Age 75+           1.700  0.190          0 5.475 3.772 7.945

  df:13     Dev:14.309    %(0):89.052   #iter:9    RSq: 0.969
```

TABLE 6.5: Logistic Regression: West London Survey of Drug-takers

According to the general logistic model given by Equation (5.6), the probability of the outcome event when all the predictors assume the value 0 is $1/(1 + e^{-a})$. From the printout in Table 6.5, an estimate of a is -4.005 giving $1/(1 + e^{4.005}) = 0.018$ for the probability. Given the way the predictor variables are coded, the group having zero values of all predictors (the referent group) comprises males with low GHQ scores aged 16–29. It follows that, for this target population, the estimated probability that a 16–19 year-old male with low GHQ score is a drug-taker is 0.018.

Estimated probabilities are obtained in the same way for the other categories. Since the regression coefficients are all positive, all the probabilities for the other groups will be greater than 0.018. They are listed below in Table 6.6. Prevalence of drug-taking increases with age and GHQ score, and females have higher rates than males.

	Age Group				
	16–29	30–44	45–64	65–74	75+
Male, low GHQ	0.018	0.038	0.063	0.094	0.091
Female, low GHQ	0.033	0.069	0.112	0.162	0.157
Male, high GHQ	0.070	0.139	0.218	0.298	0.291
Female, high GHQ	0.123	0.232	0.343	0.443	0.435

TABLE 6.6: Estimated Risks of being a Drug-taker: West London Survey

Screening using Logistic Regression

Screening is a health intervention aimed at detecting disease at an earlier stage than usual in a population at risk. For example, mammography can detect small breast cancer tumours, which can then be treated surgically before they grow or spread and require more radical treatment. However it may not be cost-effective to screen all women in the population, but rather to concentrate on higher-risk groups, and to do this you need a model, such as logistic regression, specifying the risks in the various groups.

The decision as to how high the risk needs to be before screening is justified depends on various factors, including the cost of screening and the cost of failing to detect a case that screening would have detected. For a serious disease like breast cancer, this risk may not need to be very high, say 0.01.

Based on the data analysed in Table 6.5, suppose you wish to routinely identify psychotropic drug-takers in a community using the GHQ questionnaire as an initial screening instrument, and you have the resources to properly diagnose subjects with prior likelihood greater than 0.1. Based on the estimated probabilities given in Table 6.6, this would mean properly diagnosing only males aged 45 or more with high GHQ scores, but all females except those aged less than 30 with low GHQ scores.

Many applications of this method occur in modern medicine. Based on the results of transplant operations at a particular hospital, a logistic regression model could be used to estimate the probabilities of success of alternative operations given a patient's combination of risk factors, and thus give the best advice to the patient.

Identifying Selection Bias

Logistic regression may be used to detect selection bias in epidemiologic studies. For example, a sample may be unrepresentative of a target population, or treatment and control groups being compared in an experimental study may not be balanced with respect to baseline risk factors. If you define a dichotomous variable to indicate membership of the sample (rather than the population) or the treatment group (rather than the control group), logistic regression may be used to jointly compare the baseline risk factors in the two groups.

The data in Table 6.7, cited by Thall and Vail (1990), were obtained from a clinical trial involving 59 epileptic subjects randomised to two groups receiving either the drug progabide or a placebo.

The outcome of interest in this study was the number of seizures during an eight week period after treatment, and the number of ('baseline') seizures during an eight week period prior to treatment was also recorded. To check that the treatment and control groups are balanced with respect to age and baseline seizures rates, Table 6.8 shows the printout obtained after fitting a logistic regression model with treatment group as the outcome variable and age and the number of baseline seizures as predictors. Neither age nor the baseline seizure count is statistically significant. The lower printout refers to the null model resulting from omitting these two predictors.

Since the change in deviance is only 0.599 with 2 degrees of freedom, ($p = 0.741$) you may conclude that the treatment groups are balanced. (Of course it would be surprising if the groups were not similar, since randomisation was used for the allocation.)

Care is needed when interpreting logistic regression results involving predictors that are correlated with each other. Mutually correlated predictors can mask each other, as the next example shows. The data shown comprise two sets of five measurements (X_1 = greatest length, X_2 = greatest horizontal breadth, X_3 = overall height, X_4 = upper face height, and X_5 = face breadth, all measured in millimetres), on ancient Tibetan skulls reported by Morant (1923).

| Treatment Group | | | Control Group | | |
| Patient | Number of Seizures | | Patient | Number of Seizures | |
ID Age	Outcome	Baseline	ID Age	Outcome	Baseline
1 31	14	11	29 18	42	76
2 30	14	11	30 32	28	38
3 25	11	6	31 20	7	19
4 36	13	8	32 30	13	10
5 22	55	66	33 18	19	19
6 29	22	27	34 24	11	24
7 31	12	12	35 30	74	31
8 42	95	52	36 35	20	14
9 37	22	23	37 27	10	11
10 28	33	10	38 20	24	67
11 36	66	52	39 22	29	41
12 24	30	33	40 28	4	7
13 23	16	18	41 23	6	22
14 36	42	42	42 40	12	13
15 26	59	87	43 33	65	46
16 26	16	50	44 21	26	36
17 28	6	18	45 35	39	38
18 31	123	111	46 25	7	7
19 32	15	18	47 26	32	36
20 21	16	20	48 25	3	11
21 29	14	12	49 22	302	151
22 21	14	9	50 32	13	22
23 32	13	17	51 25	26	41
24 25	30	28	52 35	10	32
25 30	143	55	53 21	70	56
26 40	6	9	54 41	13	24
27 19	10	10	55 32	15	16
28 22	53	47	56 26	51	22
			57 21	6	25
			58 36	0	13
			59 37	10	12

TABLE 6.7: Comparison of Randomisation Groups: Epileptic Drug Trial

```
Logistic Regression Analysis
Response: Allocated to Treatment Group
 Col Name          Coeff  StErr p-value  Odds    95%     CI

   0 Constant    1.0344  1.329   0.440
   3 Age        -0.0326  0.043   0.448 0.968 0.890 1.053
   5 Baseline   -0.0003  0.010   0.978 0.999 0.980 1.020

df:56   Dev:81.040   %(0):47.458   #iter:9   RSq: 0.007

 Col Name          Coeff   StErr p-value  Odds    95%     CI

   0 Constant    0.102   0.261   0.696

df:58     Dev:81.639     %(0):47.458     #iter:4     RSq: 0
```

TABLE 6.8: Comparison of Randomisation Groups: Epileptic Drug Trial

Logistic regression may be used to compare these multivariate samples, and gives the printout shown in Table 6.9. According to this printout, there are no evident differences between the skulls from the two sites, since none of the p-values is small. However if you include just the upper face height (X_4) as a predictor the picture changes, as Table 6.10 shows.

Sikkim Area						Lhasa District			
X_1	X_2	X_3	X_4	X_5		X_1	X_2	X_3	X_4	X_5
190.5	152.5	145.0	73.5	136.5		182.5	136.0	138.5	76.0	134.0
172.5	132.0	125.5	63.0	121.0		179.5	135.0	128.5	74.0	132.0
167.0	130.0	125.5	69.5	119.5		191.0	140.5	140.5	72.5	131.5
169.5	150.5	133.5	64.5	128.0		184.5	141.5	134.5	76.5	141.5
175.0	138.5	126.0	77.5	135.5		181.0	142.0	132.5	79.0	136.5
177.5	142.5	142.5	71.5	131.0		173.5	136.5	126.0	71.5	136.5
179.5	142.5	127.5	70.5	134.5		188.5	130.0	143.0	79.5	136.0
179.5	138.0	133.5	73.5	132.5		175.0	153.0	130.0	76.5	142.0
173.5	135.5	130.5	70.0	133.5		196.0	142.5	123.5	76.0	134.0
162.5	139.0	131.0	62.0	126.0		200.0	139.5	143.5	82.5	146.0
178.5	135.0	136.0	71.0	124.0		185.0	134.5	140.0	81.5	137.0
171.5	148.5	132.5	65.0	146.5		174.5	143.5	132.5	74.0	136.5
180.5	139.0	132.0	74.5	134.5		195.5	144.0	138.5	78.5	144.0
183.0	149.0	121.5	76.5	142.0		197.0	131.5	135.0	80.5	139.0
169.5	130.0	131.0	68.0	119.0		182.5	131.0	135.0	68.5	136.0
172.0	140.0	136.0	70.5	133.5						
170.0	126.5	134.5	66.0	118.5						

```
Logistic Regression Analysis
Response: Sikkim
Col Name        Coeff    StErr  p-value  Odds    95%    CI
```

Col	Name	Coeff	StErr	p-value	Odds	95%	CI
0	Constant	73.517	44.819	0.101			
2	X1	-0.129	0.097	0.183	0.879	0.727	1.063
3	X2	0.323	0.181	0.075	1.382	0.968	1.972
4	X3	-0.104	0.123	0.397	0.901	0.708	1.147
5	X4	-0.335	0.231	0.147	0.715	0.454	1.125
6	X5	-0.424	0.273	0.120	0.654	0.383	1.117

```
df:26    Dev:20.162   %(0):46.875   #iter:15   RSq: 0.544
```

TABLE 6.9: Comparison of Tibetan Skulls from Sikkim and Llasa

```
Logistic Regression Analysis
Response: Sikkim
Col Name        Coeff    StErr  p-value  Odds    95%    CI
```

Col	Name	Coeff	StErr	p-value	Odds	95%	CI
0	Constant	27.991	9.894	0.005			
5	X4	-0.380	0.134	0.005	0.684	0.526	0.89

```
df:30    Dev:28.763   %(0):46.875   #iter:13   RSq: 0.35
```

TABLE 6.10: Comparison of Tibetan Skulls: Upper Face Heights only

Some of the correlation coefficients between the skull measurements are moderately high, as the following table shows:

	X_1	X_2	X_3	X_4	X_5
X_1	1	0.11	0.43	0.76	0.57
X_2	0.11	1	0.01	0.08	0.62
X_3	0.43	0.01	1	0.29	0.20
X_4	0.76	0.08	0.29	1	0.62
X_5	0.57	0.55	0.20	0.62	1

A lesson from this example is that decisions on the basis of p-values calculated after adjusting for correlated covariates can be misleading: it is preferable to use a reduced set of uncorrelated predictors.

4: Poisson Regression

Logistic regression may be used to model data from any epidemiologic study in which the outcome variable is dichotomous: these include cross-sectional, case-control, and cohort studies. However, if the precise number of non-outcomes is not known exactly, as is the case for a cohort study using person-times instead of counts in the denominators, a method called *Poisson regression* analysis should be used.

The Poisson regression model arises as the limiting case of logistic regression when the probability that an individual subject experiences an outcome is small. To compensate for these small probabilities, the number of subjects at risk must be large. The parameter of interest is the *incidence density*, which is a measure of the risk of a new incidence of the outcome per unit person-time of exposure. This parameter will depend on the exposure variable and other predictor variables of interest, just as the probability of an outcome in the logistic model depends on these risk factors.

If there are p predictor variables x_1, x_2, ..., x_p, the Poisson regression model thus takes the form

$$\lambda = \exp(a + \sum_{j=1}^{p} b_j x_j) \tag{6.1}$$

where λ is the incidence density and a and b_j ($j = 1, 2, .., p$) are parameters similar to those in Equation (5.6).

The data layout required when fitting a Poisson regression model is precisely the same as for logistic regression with grouped data, with the person-times replacing the totals in each data record. As an illustration involving data from a simple 2-by-2 table, consider again the hypothetical example given by Breslow and Day (1987, page 95), where 14 and 5 bladder cancer deaths were assumed to have occurred among exposed and unexposed members of an industrial cohort, compared with expected numbers based on vital statistics of 5.5 and 7.3, respectively. The data layout is as follows, with the printout obtained after fitting the Poisson model to these data shown in Table 6.11.

y	L	E
14	5.5	1
5	7.3	0

```
Poisson Regression Analysis
Response: Bladder cancer
 Col Name       Coeff StErr p-value   IDR      95%    CI

   0 Constant -0.378 0.446   0.396
   3 exposed   1.313 0.519   0.011 3.716   1.343 10.282

   df:0        Dev:     0                #iter:6 RSq:1
```

TABLE 6.11: Printout from Simple Poisson Regression Model

The printout shown in Table 6.11 is similar to that given in Table 5.1 for a simple logistic regression fitted to a two-by-two table, with the *incidence density ratio* (IDR) replacing the odds ratio. The estimate for this parameter, 3.716, agrees with that obtained directly from Equation (4.13). The standard error of the natural logarithm of this estimate, 0.519, is also compatible with that given by Equation (4.14), as is the p-value for testing the null hypothesis.

Consider now the more complex data set from the Doll–Hill (1966) cohort study of British male doctors, previously discussed in Section 5 of Chapter 4. The data layout is shown below in Table 6.12.

First consider a model in which age group is a stratification variable and there is a single parameter for the effect of the exposure variable, smoking. The printout in Table 6.13 is obtained after fitting this model.

The model assumes that age group is purely a confounder, so that the relative risk (or incidence density ratio) for smokers compared with non-smokers is the same for all age groups. The estimate given in the printout for this parameter is 1.426, with a 95% confidence interval (1.155, 1.759). The estimate is very close to the Mantel–Haenszel estimate of 1.425 based on Equation (4.13).

Dead	Person-yrs	Smoker	Age Group				
			35-44	45-54	55-64	65-74	75-84
32	52 407	1	1	0	0	0	0
2	18 790	0	1	0	0	0	0
104	43 248	1	0	1	0	0	0
12	10 673	0	0	1	0	0	0
206	28 612	1	0	0	1	0	0
28	5 712	0	0	0	1	0	0
186	12 663	1	0	0	0	1	0
28	2 585	0	0	0	0	1	0
102	5 317	1	0	0	0	0	1
31	1 462	0	0	0	0	0	1

TABLE 6.12: Data Layout for Poisson Regression: British Doctors

```
Poisson Regression Analysis
Response: Coronary Deaths
Col Name          Coeff StErr p-value     IDR     95%    CI
```

Col	Name	Coeff	StErr	p-value	IDR	95%	CI
0	Constant	-7.919	0.192	0			
3	Smoker	0.355	0.107	0	1.426	1.155	1.759
5	Age 45-54	1.484	0.195	0	4.411	3.009	6.464
6	Age 55-64	2.628	0.184	0	13.839	9.655	19.836
7	Age 65-74	3.350	0.185	0	28.517	19.854	40.959
8	Age 75-84	3.700	0.192	0	40.451	27.756	58.952

```
df:4              Dev: 12.132      #iter:12     RSq:0.987
```

TABLE 6.13: Poisson Regression Model: British Doctors

The deviance gives a measure of the goodness-of-fit (provided the predicted counts in the cells are sufficiently large). The value of 12.13, with 4 degrees of freedom ($p = 0.016$) indicates a poor fit, even though the r-squared (0.987) is high.

As for logistic regression, r-statistics based on residuals may be calculated for each cell. If the model is correct, the observed counts in the various cells should be Poisson-distributed, in which case an estimate of the standard deviation of any count is given by the square root of the predicted count based on the model. These predicted counts based on the above model, together with the corresponding z-statistics, are given in Table 6.14. (A p-value for assessing the adequacy of the fitted model could be based on the sum of squares of the z-statistics, which is 11.15.)

The largest z-statistic (2.05) arises from the non-smokers in the age group 75–84, but this is not particularly large.

A model containing additional parameters could be fitted by

allowing the age groups to have different relative risks (so that age becomes an effect modifier rather than a simple confounding variable). In this case the data array would contain extra columns comprising interaction terms, or products of the exposure variable and each age-group indicator variable. The printout shown in Table 6.15 is then obtained.

Observed	Predicted	z	Smoking	Age Group
32	27.19	0.92	1	35-44
2	6.84	−1.85	0	35-44
104	98.88	0.52	1	45-54
12	17.12	−1.24	0	45-54
206	205.26	0.05	1	55-64
28	28.74	−0.14	0	55-64
186	187.19	−0.09	1	65-74
28	26.81	0.23	0	65-74
102	111.49	−0.89	1	75-84
31	21.51	2.05	0	75-84

TABLE 6.14: Predicted Values and Residuals: British Doctors

```
Poisson Regression Analysis
Response: Coronary Deaths
 Col Name         Coeff StErr p-value    IDR     95%     CI

   0 Constant   -9.148 0.707  0.000
   9 Sm.35-44    1.747 0.729  0.017   5.737   1.375 23.937
  10 Sm.45-54    0.760 0.305  0.013   2.139   1.177  3.888
  11 Sm.55-64    0.384 0.201  0.057   1.468   0.989  2.179
  12 Sm.65-74    0.305 0.203  0.133   1.356   0.911  2.018
  13 Sm.75.84   -0.100 0.205  0.625   0.905   0.605  1.352
   5 Age 45-54   2.357 0.764  0.002  10.563   2.364 47.196
   6 Age 55-64   3.830 0.732  0.000  46.070  10.975 193.39
   7 Age 65-74   4.623 0.732  0.000 101.76  24.243 427.17
   8 Age 75-84   5.294 0.730  0.000 199.21  47.678 832.35

   df:0         Dev: 0              #iter:16         RSq:1
```

TABLE 6.15: Saturated Poisson Regression Model: British Doctors

The incidence density ratios for the effect of smoking in the various age groups given in Table 6.15 are the same as those calculated directly from the data, using Equation (4.11). Note that the deviance from the model is 0: this is the *saturated* model, which fits the data exactly.

A model which fits the data exactly is of questionable usefulness. In this case the number of degrees of freedom is also 0, so there is no margin for error. It would be better to fit a model having a reasonable number of degrees of freedom.

One way of reducing the number of parameters in the model, and thus increasing the number of degrees of freedom, is to allow the coefficients defining the relative risk to depend linearly on age: this can be done simply by including an extra predictor – the interaction between smoking and age (coded as 0 for smokers aged 35–44 years, 1 for smokers aged 45–54, 2 for smokers aged 55–64, 3 for smokers aged 65–74, and 4 for smokers aged 75–84 years) – in the model shown in Table 6.14. Table 6.16 shows this printout.

The deviance is now 1.546; with 3 degrees of freedom the corresponding p-value is 0.672, so the fit is quite acceptable. Comparing Tables 6.14 and 6.16 you can see that the reduction in deviance is $12.132 - 1.546 = 11.586$. A similar result was obtained when these data were analysed in Section 5 of Chapter 4.

```
Poisson Regression Analysis
Response: Coronary Deaths
  Col Name        Coeff StErr p-value     IDR      95%     CI
```

Col	Name	Coeff	StErr	p-value	IDR	95%	CI
0	Constant	-8.586	0.305	0.000			
3	Smoker	1.136	0.281	0.000	3.115	1.795	5.405
4	Sm.age-gp	-0.309	0.097	0.002	0.734	0.607	0.889
6	Age 45-54	1.735	0.213	0.000	5.666	3.731	8.606
7	Age 55-64	3.148	0.253	0.000	23.297	14.179	38.278
8	Age 65-74	4.142	0.319	0.000	62.927	33.658	117.65
9	Age 75-84	4.731	0.384	0.000	113.40	53.430	240.68

```
  df:3              Dev: 1.546         #iter:13        RSq:0.998
```

TABLE 6.16: Model with Linear Age Effect: British Doctors

The relative risk is 3.115 (95% confidence interval 1.795 to 5.405) for the lowest age group, decreasing by a factor of 0.734 for each successive age group. According to this model the relative risk of death associated with smoking is close to 1 in age group 65–74 and slightly below 1 in age group 75–84.

Poisson regression is thus a method for analysing data from cohort studies in which the exposure times vary for different individuals or groups. Here the outcome variable for each individual is dichotomous; the Poisson distribution arises by aggregating outcomes from many individuals in the same exposure group.

Accidents resulting in wave damage to cargo-carrying vessels provide another illustration of Poisson regression analysis. The data listed in Table 6.17 are taken from McCullagh and Nelder (1989, page 205) and were provided by J. Crilley and L.N. Heminway of Lloyd's

Register of Shipping. The records in this data table correspond to groups of ships of the same type, five-year construction period, and period of operation, so that only the aggregate numbers of damage incidences are given, together with the aggregated duration of exposure in months.

Number of accidents	Service period	Ship type	Period built	Operation period
0	127	A	1960-64	1960-74
0	63	A	1960-64	1975-79
3	1095	A	1965-69	1960-74
4	1095	A	1965-69	1975-79
6	1512	A	1970-74	1960-74
18	3353	A	1970-74	1975-79
11	2244	A	1975-79	1975-79
39	44882	B	1960-64	1960-74
29	17176	B	1960-64	1975-79
58	28609	B	1965-69	1960-74
53	20370	B	1965-69	1975-79
12	7064	B	1970-74	1960-74
44	13099	B	1970-74	1975-79
18	7117	B	1975-79	1975-79
1	1179	C	1960-64	1960-74
1	552	C	1960-64	1975-79
0	781	C	1965-69	1960-74
1	676	C	1965-69	1975-79
6	783	C	1970-74	1960-74
2	1948	C	1970-74	1975-79
1	274	C	1975-79	1975-79
0	251	D	1960-64	1960-74
0	105	D	1960-64	1975-79
0	288	D	1965-69	1960-74
0	192	D	1965-69	1975-79
2	349	D	1970-74	1960-74
11	1208	D	1970-74	1975-79
4	2051	D	1975-79	1975-79
0	45	E	1960-64	1960-74
7	789	E	1965-69	1960-74
7	437	E	1965-69	1975-79
5	1157	E	1970-74	1960-74
12	2161	E	1970-74	1975-79
1	542	E	1975-79	1975-79

TABLE 6.17: Numbers of Shipping Accidents

Table 6.18 shows the printout after fitting a Poisson model. This model includes effects for the three risk factors but no interactions. The deviance from this model is 38.695 with 25 degrees of freedom, giving a p-value of 0.04 which indicates that the fit of the model is inadequate.

The model may be assessed further by graphing standardised

residuals (obtained as described earlier by dividing the residual cell count by the square root of the predicted count) against normal scores, and these are shown in Figure 6.1.

```
Poisson Regression Analysis
Response: Damage
 Col Name              Coeff StErr p-value   IDR   95%    CI

   0 Constant         -6.406 0.217  0.000
   4 Type  B          -0.543 0.177  0.002 0.581  0.410 0.822
   5 Type  C          -0.687 0.329  0.037 0.503  0.264 0.958
   6 Type  D          -0.076 0.290  0.794 0.927  0.525 1.638
   7 Type  E           0.326 0.236  0.167 1.385  0.873 2.198
   9 Built 65-69       0.697 0.149  0.000 2.008  1.498 2.692
  10 Built 70-74       0.818 0.170  0.000 2.267  1.626 3.161
  11 Built 75-79       0.453 0.233  0.052 1.574  0.997 2.484
  12 Period 75-79      0.384 0.118  0.001 1.469  1.165 1.852

 df:25        Dev: 38.695        #iter:9        RSq:0.736
```

TABLE 6.18: Poisson Model: Ship Accidents

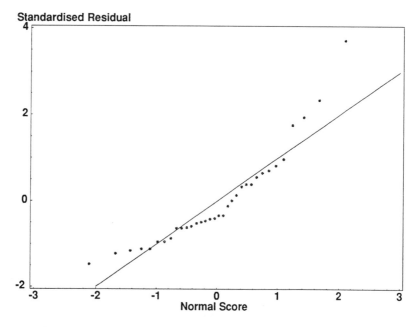

FIGURE 6.1: Standardised Residuals from Poisson Model: Ship Accidents

If the Poisson model were correct the standardised residuals should cluster around the straight line with slope 1, which is superimposed on the graph in Figure 6.1. You can see that the right-most residual, which corresponds to ships of type C built in the period 1970–74 and operating during the period 1960–74, is a clear outlier: the model predicts 1.47 accidents whereas 6 were observed. Table 6.19 shows the refitted model with this outlier omitted.

```
Poisson Regression Analysis
Response: Damage
 Col Name          Coeff StErr p-value  IDR      95%   CI

   0 Constant     -6.406 0.218  0.000
   4 Type B       -0.555 0.177  0.002 0.574  0.405 0.813
   5 Type C       -1.260 0.438  0.004 0.284  0.120 0.669
   6 Type D       -0.080 0.290  0.784 0.923  0.523 1.632
   7 Type E        0.332 0.236  0.159 1.394  0.878 2.213
   9 Built 65-69   0.686 0.150  0.000 1.986  1.481 2.662
  10 Built 70-74   0.760 0.172  0.000 2.137  1.527 2.992
  11 Built 75-79   0.407 0.233  0.081 1.502  0.951 2.373
  12 Period 75-79  0.443 0.120  0.000 1.557  1.231 1.969

  df:24        Dev: 28.286      #iter:9      RSq:0.798
```

TABLE 6.19: Poisson Model with one Outlier omitted: Ship Accidents

The deviance is now 28.286 with 24 degrees of freedom, giving a goodness-of-fit p-value of 0.25, so the fit is adequate.

The interpretation of the model is straightforward. Ships of type B and C are had lower risk of suffering damage incidents, as did those built in the years 1960 to 1964, while those operating during the period from 1975 to 1979 were more accident-prone.

Poisson Regression with Case-by-Case Data

Poisson regression may also be used to analyse data where the outcome variable for an individual is the total number of events experienced during some exposure period. The New South Wales Water Board, concerned about pollution in beaches around Sydney due to sewage effluent, undertook a pilot study in 1990–91 which involved monitoring ear infections among beach swimmers and non-beach (mostly pool) swimmers. The data are listed in Table 6.20.

In this study the outcome variable is the number of ear infections reported by the subject during the season. There were 287 subjects in the study classified by frequent or infrequent swimmers, location of swimming (beach or other), age group (15–19, 20–24 or 25–29) and gender. The printout from fitting the Poisson regression model to the case-by-case data is shown in Table 6.21. The referent group is taken to be infrequent male beach swimmers aged 25–29.

Freq	Location	Sex	Age	Outcomes
freq	other	m	15–19 :	0 0 0 0 0 0 0 0 0 0 0 1 1 1 1 1 1 1 2 2
				2 2 3 3 3 3 4 4 6 11 16
freq	other	f	15–19 :	0 0 4 10
freq	other	m	20–24 :	0 0 0 0 1 2 2 2 3 3 5 17
freq	other	f	20–24 :	0 0 0 0 0 1 2 3 3 3 4
freq	other	m	25–29 :	1 1 2 3 4 4 10
freq	other	f	25–29 :	0 0 2 2
freq	beach	m	15–19 :	0 0 0 0 0 0 0 0 0 0 1 1 2 2 3 4 4 4 9
freq	beach	f	15–19 :	0 0 0 0 0 0 0 0 1 2 3 3 6 9
freq	beach	m	20–24 :	0 0 0 0 0 0 1
freq	beach	f	20–24 :	0 0 0 0 0 1 1 2 3
freq	beach	m	25–29 :	0 0 0 0 0 0 0 0 0 0 0 1 1 2 2 9
freq	beach	f	25–29 :	0 0 0 1 3 4 5 6
infrq	other	m	15–19 :	0 0 0 0 0 0 0 0 0 0 0 0 0 0 0 1 1 1 2 2
				2 2 2 3 3 3 3 4 4 4 5
infrq	other	f	15–19 :	0 0 0 6
infrq	other	m	20–24 :	0 0 0 0 0 0 0 1 1 2 2 3 3
infrq	other	f	20–24 :	0 0 0 0 0 1 1 2 2 3
infrq	other	m	25–29 :	0 0 0 0 1 1 3 3
infrq	other	f	25–29 :	0 0 0 1
infrq	beach	m	15–19 :	0 0 0 0 0 0 0 0 0 0 0 0 1 1 1 2 2 2 2 5
infrq	beach	f	15–19 :	0 0 0 0 0 0 0 1 2 2 2 3 10
infrq	beach	m	20–24 :	0 0 0 0 1 1 5
infrq	beach	f	20–24 :	0 0 0 0 0 0 1 1 1 2
infrq	beach	m	25–29 :	0 0 0 0 0 0 0 0 0 0 0 0 1 2 2
infrq	beach	f	25–29 :	0 0 0 2 2 2

TABLE 6.20: Ear Infections among 287 Swimmers

```
Poisson Regression Analysis
Response: Number of Ear Infections
 Col Name         Coeff  StErr p-value    IDR     95%      CI
```

Col	Name	Coeff	StErr	p-value	IDR	95%	CI
0	Constant	-0.402	0.146	0.006			
2	Frequent	0.611	0.105	0.000	1.843	1.501	2.263
3	Non-beach	0.535	0.106	0.000	1.707	1.385	2.103
4	Female	0.090	0.112	0.423	1.094	0.878	1.363
5	15-19	0.190	0.130	0.143	1.209	0.938	1.558
6	20-24	-0.185	0.155	0.233	0.831	0.614	1.126

```
 df:281           Dev: 755.433        #iter:8      RSq:0.084
```

TABLE 6.21: Poisson Model fitted to Case-by-case Data: Ear Infections

The case-by-case data layout comprises 287 records, but this may be reduced to 24 records by aggregating outcomes (Table 6.22).

Fitting the Poisson regression model to the grouped data now gives the printout in Table 6.23.

Count	Total	Frequency	Location	Sex	Age
68	31	freq	other	m	15-19
14	4	freq	other	f	15-19
35	12	freq	other	m	20-24
16	11	freq	other	f	20-24
25	7	freq	other	m	25-29
4	4	freq	other	f	25-29
30	20	freq	beach	m	15-19
24	15	freq	beach	f	15-19
1	7	freq	beach	m	20-24
7	9	freq	beach	f	20-24
15	16	freq	beach	m	25-29
19	8	freq	beach	f	25-29
45	32	infrq	other	m	15-19
6	4	infrq	other	f	15-19
12	13	infrq	other	m	20-24
9	10	infrq	other	f	20-24
8	8	infrq	other	m	25-29
1	4	infrq	other	f	25-29
16	20	infrq	beach	m	15-19
20	14	infrq	beach	f	15-19
7	7	infrq	beach	m	20-24
5	10	infrq	beach	f	20-24
5	15	infrq	beach	m	25-29
6	6	infrq	beach	f	25-29

TABLE 6.22: Aggregated data from Rable 6.20

```
Poisson Regression Analysis
Response: Number of Ear Infections
Col Name         Coeff  StErr p-value    IDR     95%    CI

  0 Constant    -0.402  0.146   0.006
  3 Frequent     0.611  0.105   0.000  1.843  1.501  2.263
  4 Non-beach    0.535  0.106   0.000  1.707  1.385  2.103
  5 Female       0.090  0.112   0.423  1.094  0.878  1.363
  6 15-19        0.190  0.130   0.143  1.209  0.938  1.558
  7 20-24       -0.185  0.155   0.233  0.831  0.614  1.126

   df:18          Dev: 51.714        #iter:8    RSq:0.572
```

TABLE 6.23: Poisson Regression with Grouped Data: Ear Infections

The deviance in the printout in Table 6.23 is high compared to the number of degrees of freedom, indicating a poor fit ($p < 0.0005$). The standardised residuals are plotted against normal scores in Figure

6.2, with the unit slope line superimposed.

There are two standardised residuals greater than 2.5 in magnitude, corresponding to the frequent male non-beach swimmers aged 20–24 and 25–29, respectively. Looking at the data in the case-by-case layout, you can see that there are two swimmers with unusually high numbers of infections (17 and 10, respectively). However with these outliers omitted, the deviance (32.06) is still too high ($p = 0.022$).

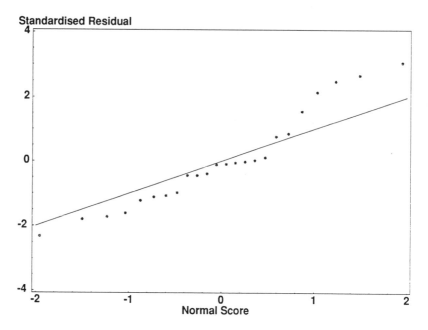

FIGURE 6.2: Standardised Residuals from Poisson Model: Ear Infections

If the deviance from a fitted Poisson model is too high, there is said to be *overdispersion*, or *extra-Poisson variation*, indicating that there is more variability in the outcomes than can be accounted for by the implicit assumption that the outcomes for different subjects are independent. In the present study, it could be that some swimmers are particularly prone to ear infections due to other risk factors besides those measured in the study.

While it appears from our analysis that more frequent swimmers and non-beach swimmers are at higher risk of ear infection, the poor fit of the model reduces the evidence for this conclusion.

5: Modelling Outcome Severity

In the study of ear infections among swimmers the outcome variable for a subject takes non-negative integer values, and is a measure of disease severity, in the sense that a subject with two ear infections is affected more severely than a subject with just one infection, and a subject with three infections has an even more severe level of disease.

In the Poisson model, outcomes are interchangeable in the sense that, for a group of subjects with the same combination of risk factors, only the total number of outcomes is important. In the swimmers study, for example, you can see from the last record in Table 6.22 that there were six 25–29 year-old female infrequent beach swimmers, and they suffered a total of 6 ear infections between them: three had no infections and the other three each had two infections. You would get the same result from the Poisson model if, say, one of the six had suffered 6 infections and the other five had been infection-free.

However, you would get a different result if the outcomes were differently weighted. Suppose you said that the precise number of infections suffered by a swimmer is not important, and all that matters is whether the swimmer was infected or not. Then you could fit a logistic regression model to the data, obtaining the result shown in Table 6.24. The deviance from this model is close to the number of degrees of freedom so the fit appears to be quite satisfactory. The result is similar to that obtained from the Poisson model (Table 6.23), but the effect of frequent swimming is no longer statistically significant.

```
Logistic Regression Analysis
Response: Ear Infection
 Col Name          Coeff   StErr  p-value  Odds    95%    CI
```

Col	Name	Coeff	StErr	p-value	Odds	95%	CI
0	Constant	-0.711	0.311	0.022			
3	Frequent	0.404	0.242	0.095	1.499	0.933	2.408
4	Non-beach	0.738	0.251	0.003	2.093	1.281	3.420
5	Female	0.030	0.265	0.911	1.030	0.613	1.732
6	15-19	0.096	0.306	0.753	1.101	0.605	2.004
7	20-24	-0.068	0.350	0.847	0.935	0.471	1.855

```
df:18    Dev:18.271   %(0):52.613   #iter:8   RSq: 0.399
```

TABLE 6.24: Logistic Regression: Ear Infections

Information is lost by ignoring the severity of the disease. The severity information could be retained to some extent by coding the outcomes in sets of three (infection 'outbreaks') as in the following list. Recoding the outcomes in this way reduces the variance in the distribution of outcomes while still differentiating between subjects who experienced a greater severity of infection. When the Poisson regression model is now fitted to the recoded data there is no overdispersion, as the following printout (Table 6.25) shows.

no infections:	0	10–13 infections:	4
1–3 infections:	1	14–16 infections:	5
4–6 infections:	2	17+ infections:	6
7–9 infections:	3		

```
Poisson Regression Analysis
Response: Ear Infection Severity
  Col Name         Coeff  StErr p-value    IDR    95%     CI
```

Col	Name	Coeff	StErr	p-value	IDR	95%	CI
0	Constant	-0.996	0.207	0.000			
3	Frequent	0.482	0.151	0.001	1.620	1.206	2.176
4	Non-beach	0.463	0.155	0.003	1.588	1.173	2.151
5	Female	0.028	0.165	0.864	1.028	0.745	1.420
6	15-19	0.132	0.188	0.483	1.141	0.789	1.651
7	20-24	-0.144	0.223	0.519	0.866	0.560	1.340

```
df:18        Dev: 18.494        #iter:8        RSq:0.541
```

TABLE 6.25: Poisson Regression: Ear Infection Severity

When fitting a Poisson model to grouped data, the deviance gives an overall measure of goodness-of-fit, and plotting standardised residuals against normal scores can identify discrepant groups. If outcomes for individual subjects are' available it is also possible to see how well the Poisson distribution matches the histogram of the outcomes. For a given incidence density parameter λ the Poisson distribution has a particular shape given by the formula

$$Prob[Y = k] = \frac{\lambda^k}{k!}e^{-\lambda} \tag{6.2}$$

This distribution also has the property that if Y_1 and Y_2 have independent Poisson distributions with parameters λ_1 and λ_2 respectively, their sum Y_1+Y_2 has a Poisson distribution too, with parameter $\lambda_1+\lambda_2$.

Figure 6.3 shows histograms of the outcomes (on the left) and the coded outcomes (on the right) for the 287 subjects who participated in the swimmers study, with fitted Poisson distributions (represented by the filled bars). The incidence density parameter is estimated in each case by dividing the total number of outcomes (or coded outcomes) by the number of subjects, giving estimates 398/287 = 1.39 and 186/287 = 0.65, respectively. It is clear that the fit of the Poisson model to the numbers of infections is poor but is substantially improved when the outcomes are recoded. The fit may be improved further by a slightly different grouping of the outcome categories.

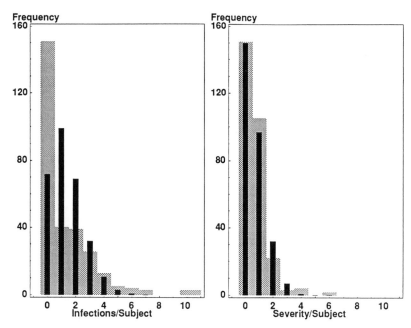

FIGURE 6.3: Poisson Models for Ear Infections (recoded in right panel)

As a further illustration of the method, consider some social class data collected from fathers and their sons, from Payne (1992).

| | | \multicolumn{9}{c}{Son's Social Class} |
		1	2	3	4	5	6	7	8	9
Father's	1	1	0	3	7	2	1	0	0	0
Social	2	4	20	9	22	16	2	0	0	0
Class	3	4	12	5	7	5	2	0	0	0
	4	12	54	32	143	50	10	2	0	2
	5	5	29	34	116	67	22	4	0	2
	6	0	8	8	50	18	12	0	0	0
	7	0	2	0	1	1	1	1	0	0
	8	3	1	0	3	3	1	0	0	0
	9	1	1	6	10	7	2	1	0	0

TABLE 6.26: Social Classes of Father–Son Pairs

One approach to these data involves recoding the son's social class to approximate a Poisson distribution and then fitting a Poisson regression model with father's social class as a predictor variable. The printout in Table 6.27 is obtained if the following recoding is used for both son's and father's social class.

Social class 1–3: 0
Social class 4:   1
Social class 5:   2
Social class 6:   3
Social class 7–9: 4

```
Poisson Regression Analysis
Response: Son's Social Class (recoded)
  Col Name        Coeff StErr p-value    IDR      95%   CI
```

Col	Name	Coeff	StErr	p-value	IDR	95%	CI
0	Constant	-0.229	0.019	0			
4	Father=4	0.175	0.033	0	1.192	1.117	1.271
5	Father=5	0.427	0.030	0	1.533	1.444	1.627
6	Father=6	0.469	0.050	0	1.598	1.448	1.764
7	Father>6	0.448	0.074	0	1.565	1.354	1.809

```
df:0              Dev: 0              #iter:11            RSq: 1
```

TABLE 6.27: Poisson Regression: Social Class Data

The result may be interpreted in terms of conditional probabilities calculated from the estimated parameters in the model. The probability of finding a son in a specified social class is given by Equation (6.2) where the parameter λ is dependent on the social class of the father and is given by Equation (6.1). For example, the probability that the son is in social class 1–3, given that the father's social class is also 1–3, is $\exp(-\exp(-0.229))$, that is, 0.451. Similarly, the probability of finding a son in Class 1–3 given that the father is in Class 4 is estimated to be $\exp(-\exp(-0.229+0.175)) = \exp(-0.947) = 0.388$, while the probability that a son is in Class 5 given that his father is in Class 4 is $\frac{1}{2}(0.947)^2\exp(-0.947) = 0.174$. Moreover father's classes 5, 6, and 7–9 have statistically indistinguishable effects on the son's social class.

The goodness-of-fit of the model may be assessed by comparing the expected frequencies based on the fitted model with those in the corresponding row of the table of observed counts. For example the second row of the (recoded) data table (father's class 4) has model-based probabilities 0.388, 0.367, 0.174, 0.055, and 0.016 corresponding to son's classes 1–3, 4, 5, 6, and 7–9, respectively, and allocating the 305 subjects according to these probabilities gives expected counts 118, 112, 53, 17 and 5, respectively. These may be compared with the observed counts 98, 143, 50, 10 and 4, respectively. Table 6.28 gives the result of all such calculations, with the expected counts enclosed in parentheses.

		Son's Social Class				
		1–3	4	5	6	7–9
Father's	1–3	58 (55)	36 (44)	23 (17)	5 (5)	0 (1)
Social	4	98 (118)	143 (112)	50 (53)	10 (17)	4 (5)
Class	5	68 (82)	116 (101)	67 (61)	22 (25)	6 (10)
	6	16 (27)	50 (34)	18 (22)	12 (9)	0 (4)
	7–9	14 (13)	14 (16)	11 (10)	4 (4)	2 (2)

TABLE 6.28: Fitted Poisson Model: Social Class Data

A p-value for the goodness-of-fit may be calculated using a chi-squared test. The differences between the observed and expected counts are squared, scaled by dividing by the expected counts, and then summed, and the result is then compared with a chi-squared distribution with appropriate degrees of freedom. If there are c outcome categories whose probabilities sum to 1, there are $c-1$ degrees of freedom in each row of the table, so the total number of degrees of freedom is $r(c-1)-p$, where r is the number of rows in the table and p is the number of estimated parameters in the model. These calculations give a chi-squared statistic 45.29 with 15 degrees of freedom, so the fit of the Poisson model is poor (p-value 0.0001).

The poor fit should not surprise you: the Poisson model (Equation (6.2)) is rather restrictive and does not have sufficient flexibility to provide a good model for general ordered categorical data. More general models for both ordinal and general categorical outcomes are considered by McCullagh and Nelder (1989) and by Agresti (1990).

6: Survival Data

The data listed in Table 6.29, reported by Hand et al (1993), arose from an investigation by Hoenig and Crotty (1958) of length of stay in a psychiatric observation ward. They involve four groups of 336 patients who left the ward to go to another ward or to another mental hospital.

Do the departure rates in the four groups differ? To answer this question, think of these data as arising from a cohort study with departure from the ward as the outcome event and patient group as the determinant. The following table (6.30) gives the total number of departures in each group during the first seven days (for which complete information is available), and the total length of stay in person-days during this period.

| | | Type of Patient | | | |
		Male voluntary	*Female voluntary*	*Male certified*	*Female certified*
	1	5	9	4	11
	2	16	25	18	18
Number of	3	20	34	20	28
days in ward	4	10	17	6	8
	5	5	15	1	12
	6	3	8	0	5
	7	3	7	0	5
	8+	5	11	1	6

TABLE 6.29: Length of Stay of Patients in a Psychiatric Ward

No. departures	Person-days	Group
62	236	Male voluntary
115	478	Female voluntary
49	136	Male certified
87	330	Female certified

TABLE 6.30: Person–days of Stay by Patient Group

Table 6.31 shows the printout obtained from Poisson regression, with certified females taken as the referent group. You can see that the certified male patients appear to have a higher departure rate than the other groups, but the difference is not statistically significant. (Since certified males are potentially violent, it would not be surprising if they had been transferred earlier than the other patients.)

```
Poisson Regression Analysis
Response: Departure from Psychiatric Ward
  Col Name        Coeff StErr p-value    IDR      95%   CI
```

Col	Name	Coeff	StErr	p-value	IDR	95%	CI
0	Constant	-1.333	0.106	0.000			
4	Group=FV	-0.092	0.141	0.516	0.913	0.693	1.203
5	Group=MC	0.312	0.177	0.078	1.367	0.965	1.935
6	Group=MV	-0.004	0.165	0.983	0.996	0.721	1.377

```
  df:0          Dev: 0        #iter:7         RSq: 1
```

TABLE 6.31: Poisson Regression: Departures from a Psychiatric Ward

A problem with this analysis is that Poisson regression assumes that the risk of an event is small. In this case the 'event' is a departure by a patient on any particular day, and the model fits different risks for each of the four patient groups, but assumes that for each group these risks of departure are the same for different days. But these estimated risks are moderately large: they vary from 115/478 = 0.22 for female voluntary patients to 49/136 = 0.36 for male certified patients.

Logistic regression, which does not require the risks to be small, is more appropriate for these data, and gives the printout shown in Table 6.32. Note that the p-value corresponding to the difference in departure rate between female certified and male certified patients is now statistically significant.

```
Logistic Regression Analysis
Response: Departure from Psychiatric Ward
   Col Name        Coeff  StErr  p-value  Odds     95%    CI

     0 Constant   -1.027  0.125   0.000
     4 Group=FV   -0.122  0.164   0.457  0.885   0.641  1.221
     5 Group=MC    0.453  0.218   0.038  1.573   1.026  2.411
     6 Group=MV   -0.005  0.194   0.980  0.995   0.681  1.454

 df:0        Dev:0      %(0):73.475     #iter:7      RSq: 1
```

TABLE 6.32: Logistic Regression: Departures from a Psychiatric Ward

This analysis is based on 1180 patient-days rather than 336 patients. In logistic regression analysis, in common with other forms of regression analysis, it is assumed that the data are independent. In the present context this means that even though there are only 336 patients the analysis is based on the 1180 departure decisions, and these are assumed to be mutually independent events.

The model also assumes that the risk of departure does not vary with day of departure, an assumption that is questionable for these data. Since the length of stay for each patient is known, you can fit a more complex model which allows the departure rate to vary with duration of stay in the ward. The following data layout (Table 6.33) includes day of departure as an additional stratification variable in the model, and the printout in Table 6.34 is then obtained after fitting the more complex logistic regression model, with Day 1 taken as the referent category. This model is considered by Agresti (1990, page 319).

The difference in departure rate for male certified and female certified patients is now statistically highly significant. In fact the effect of the adjustment for day of departure is to increase the odds ratio of departure for this group relative to the other groups, which means that day of departure is a confounder. The risk of departure is relatively lower in the first two days.

No. departures	No. at risk	Group	Departure day
5	67	MV	1
16	62	MV	2
20	46	MV	3
10	26	MV	4
5	16	MV	5
3	11	MV	6
3	8	MV	7
9	126	FV	1
25	117	FV	2
34	92	FV	3
17	58	FV	4
15	41	FV	5
8	26	FV	6
7	18	FV	7
4	50	MC	1
18	46	MC	2
20	28	MC	3
6	8	MC	4
1	2	MC	5
0	1	MC	6
0	1	MC	7
11	93	FC	1
18	82	FC	2
28	64	FC	3
8	36	FC	4
12	28	FC	5
5	16	FC	6
5	11	FC	7

TABLE 6.33: Data Layout including Day of Departure as a Covariate

```
Logistic Regression Analysis
Response: Departure from Psychiatric Ward
  Col Name         Coeff  StErr p-value  Odds   95%     CI
```

Col	Name	Coeff	StErr	p-value	Odds	95%	CI
0	Constant	-2.447	0.228	0.000			
6	Type=FV	-0.162	0.172	0.349	0.851	0.607	1.193
7	Type=MC	0.723	0.237	0.002	2.060	1.294	3.280
8	Type=MV	0.010	0.203	0.960	1.010	0.678	1.505
10	Day=2	1.284	0.236	0.000	3.611	2.272	5.739
11	Day=3	2.193	0.238	0.000	8.966	5.621	14.303
12	Day=4	1.714	0.275	0.000	5.553	3.237	9.524
13	Day=5	2.010	0.299	0.000	7.466	4.154	13.418
14	Day=6	1.641	0.360	0.000	5.160	2.548	10.450
15	Day=7	2.073	0.389	0.000	7.951	3.712	17.033

```
df:18     Dev:14.33    %(0):73.475    #iter:9    RSq: 0.899
```

TABLE 6.34: Model for Ward Departures with Day of Departure included

Table 6.35 shows what happens when the model is fitted to the data from Table 6.33 with day of departure omitted from the analysis. The parameter estimates and standard errors are the same as those obtained from the aggregated data (Table 6.34) but the high deviance (135.01 with 24 degrees of freedom) shows that the model fits very poorly.

```
Logistic Regression Analysis
Response: Departure from Psychiatric Ward
  Col Name           Coeff  StErr  p-value  Odds   95%    CI

    0 Constant       -1.027  0.125   0.000
    6 Type=FV        -0.122  0.164   0.457  0.885  0.641  1.221
    7 Type=MC         0.453  0.218   0.038  1.573  1.026  2.411
    8 Type=MV        -0.005  0.194   0.980  0.995  0.681  1.454

  df:24      Dev:135.011  %(0):73.475  #iter:7    RSq: 0.052
```

TABLE 6.35: Model for Ward Departures with Day of Departure omitted

This example provides a simple illustration of *survival analysis*. Further methods for the analysis of such data, where it is assumed that the risk of an event may depend on the duration of follow-up as well as the risk factors of interest, are discussed in the next chapter.

Summary

In this chapter the methods of logistic regression introduced in Chapter 5 are developed further, and Poisson regression is described.

Case-by-Case Data

The data layout for fitting a logistic regression model may consist of grouped data (as in the examples given in Chapter 5) or individual ('case-by-case') records. Case-by-case records are needed when a predictor has a continuous range. The results are the same except for the deviance, which can provide a measure of goodness-of-fit with grouped data but not with case-by-case data. Studying birth weight as a risk factor for perinatal mortality provides an illustration of logistic modelling with a continuously varying risk factor.

Modelling Risks

If the data have been collected from a cross-sectional study or a cohort study an estimate of the risk of outcome is directly available. Identifying higher-risk groups is useful for planning screening programs. The method is illustrated using data from a cross-sectional study of drug-takers in West London. In this section we also show how logistic regression can detect selection bias in a study.

Poisson Regression

Poisson regression is the limiting case of logistic regression when the risks of outcome tend to zero, and provides a method for analysing data from cohort studies where the exposure is measured in person-times. The method is applied to the study of coronary deaths among British doctors, previously considered in Chapter 3, to a survey of shipping accidents, and to a longitudinal study of ear infections among Sydney swimmers.

Modelling Outcome Severity

The Poisson regression model is shown to be more general than it first appears, providing a possible model for ordinal outcomes. This approach is applied (successfully) to alternative measures of disease severity of the swimmers' ear infections, and (unsuccessfully) to an investigation of social class in father–son pairs.

Survival Data

The time at which an event occurs may be a confounder in a cohort study, but logistic and Poisson regression can handle this situation. This method is applied to data from a study comparing rates of departure of various groups of patients from a psychiatric observational ward.

Exercises

Exercise 6.1: For Dr Hieu's study (see the data listed in the appendix), use logistic regression to model the risk of perinatal death as a function of both birth weight and mother's education level, and give an interpretation of your result.

Exercise 6.2: Van Vliet & Gupta (1973) reported mortality and birth weights of 50 infants with severe idiopathic respiratory distress syndrome as follows (an asterisk denotes death outcome). Fit a logistic regression model to these data, and use the model to estimate the maximum birth weight for which the risk of death is greater than 20%.

1.050*	2.500*	1.890*	1.760	2.830
1.175*	1.030*	1.940*	1.930	1.410
1.230*	1.100*	2.200*	2.015	1.715
1.310*	1.185*	2.270*	2.090	1.720
1.500*	1.225*	2.440*	2.600	2.040
1.600*	1.262*	2.560*	2.700	2.200
1.720*	1.295*	2.730*	2.950	2.400
1.750*	1.300*	1.130	3.160	2.550
1.770*	1.550*	1.575	3.400	2.570
2.275*	1.820*	1.680	3.640	3.005

Exercise 6.3: Use Poisson regression to analyse the data from the post-menopausal hormone use/coronary heart disease cohort study cited in Exercise 4.2, comparing your result with that obtained using the methods outlined in Section 4 of Chapter 4.

Exercise 6.4: Gilchrist (1984, page 132) reported the results from an insect trapping experiment, where 33 traps were set out across sand dunes and the numbers of different insects caught in a fixed time were recorded as follows:

| Taxa | Number of Individuals in a Trap | | | | | | |
	0	1	2	3	4	5	6
Staphylinoidea	10	9	5	5	1	2	1
Hemiptera	6	8	12	4	3	0	0

Fit a Poisson regression model to these data, and examine its goodness-of-fit.

Exercise 6.5: Use Poisson regression to analyse the following data:

| | Minneapolis-St.Paul | | Dallas-Fort Worth | |
Age Group	Cases	Population	Cases	Population
15-24	1	172675	4	181343
25-34	16	123065	38	146207
35-44	30	96216	119	121374
45-54	71	92051	221	111353
55-64	102	72159	259	83004
65-74	130	54722	310	55932
75-84	133	32185	226	29007
85+	40	8328	65	7538

These are the number of cases of non-melanoma skin cancer and the population size for women from the two cities Minneapolis-St.Paul and Dallas-Fort Worth in eight age groups. (The data were adapted by Kleinbaum et al (1988, Chapter 21) from J. Scotto, A.W. Kopf & F. Urbach, *Cancer*, (1974), **34**, pages 1333-1338.)

Exercise 6.6: The following data are the numbers of Prussian Militar-personen killed by horse kicks for each of 14 corps in 20 successive years. These data have been considered by statisticians as an example of Poisson-distributed data (see, for example, Preece et al, 1988). Fit a Poisson regression model and thus assess the corps and year effects.

| | | | | Corps | | | | | | | | | | |
Year	G	I	II	III	IV	V	VI	VII	VIII	IX	X	XI	XIV	XV
1875	0	0	0	0	0	0	0	1	1	0	0	0	1	0
1876	2	0	0	0	1	0	0	0	0	0	0	0	1	1
1877	2	0	0	0	0	0	1	1	0	0	1	0	2	0
1878	1	2	2	1	1	0	0	0	0	0	1	0	1	0
1879	0	0	0	1	1	2	2	0	1	0	0	2	1	0
1880	0	3	2	1	1	1	0	0	0	2	1	4	3	0
1881	1	0	0	2	1	0	0	1	0	1	0	0	0	0
1882	1	2	0	0	0	0	1	0	1	1	2	1	4	1
1883	0	0	1	2	0	1	2	1	0	1	0	3	0	0
1884	3	0	1	0	0	0	0	1	0	0	2	0	1	1
1885	0	0	0	0	0	0	1	0	0	2	0	1	0	1
1886	2	1	0	0	1	1	1	0	0	1	0	1	3	0
1887	1	1	2	1	0	0	3	2	1	1	0	1	2	0
1888	0	1	1	0	0	1	1	0	0	0	0	1	1	0
1889	0	0	1	1	0	1	1	0	0	1	2	2	0	2
1890	1	2	0	2	0	1	1	2	0	2	1	1	2	2
1891	0	0	0	1	1	1	0	1	1	0	3	3	1	0
1892	1	3	2	0	1	1	3	0	1	1	0	1	1	0
1893	0	1	0	0	0	1	0	2	0	0	1	3	0	0
1894	1	0	0	0	0	0	0	0	1	0	1	1	0	0

Exercise 6.7: Analyse the data given in Exercise 4.3 using Poisson regression. Does the risk vary with elapsed time?

References

Agresti, A. (1990): *Categorical Data Analysis*. John Wiley & Sons. New York.

Bishop, Y.M.M. (1969); Full contingency tables, logits, and split contingency tables, *Biometrics*, **25**, pages 383–399.

Breslow, N.E. & N.E. Day (1987): *Statistical Methods in Cancer Research: Volume II – The Design and Analysis of Cohort Studies*. IARC. Lyon.

Doll, R. & A.B. Hill (1966): Mortality of British doctors in relation to smoking: observations on coronary thrombosis. *Natl Cancer Inst. Monogr.*, **19**, pages 205–268.

Gilchrist, W. (1984): *Statistical Modeling*. John Wiley & Sons. Chichester.

Hand, D.J., F. Daly, A.D. Lunn, K.J. McConway & E. Ostrowski (1993): *A Handbook of Small Data Sets*. Chapman & Hall. London.

Hoenig, J. & J.M. Crotty (1958): *International Journal of Social Psychiatry*, **3**, pages 260–277.

Kleinbaum, D.G., L.L. Kupper & K.E. Muller (1988): *Applied Regression Analysis and Other Multivariate Methods* (2nd ed.). PWS-Kent. Boston.

McCullagh, P. & J.A. Nelder (1989): *Generalized Linear Models* (2nd ed.). Chapman & Hall. London.

Morant, G.M. (1923): A first study of the Tibetan skull, *Biometrika*, **14**, pages 193–260.

Murray, J.D., G. Dunn, P. Williams & A. Tarnopolsky (1981): Factors affecting the consumption of psychotropic drugs, *Psychological Medicine*, **11**, pages 551–560.

Payne, A.C. (1992): *Confounding variables and selection effects in a follow-up study*. Southampton University MSc dissertation (Table 5.9), Faculty of mathematical Statistics.

Preece, D.A., G.J.S. Ross, & S.P.J. Kirby (1988): Bortkewitsch's horse-kicks and the generalized linear model, *The Statistician*, **37**, pages 313–318.

Thall, P.F & S.C. Vail (1990): Some covariance models for longitudinal count data with overdispersion, *Biometrics*, **46**, pages 657–671.

Van Vliet, P.K. & J.M. Gupta (1973): Tham-v-sodium bicarbonate in idiopathic respiratory distress syndrome, *Archives of Disease in Childhood*, **48**, pages 249–255.

7

SURVIVAL ANALYSIS

1: Introduction

Survival analysis is concerned with measuring the risk of occurrence of an outcome event as a function of time. It thus focuses on the duration of time elapsed from when a subject enters a study until the event occurs, and uses the *survival curve* to describe its distribution. Survival analysis is also concerned with the comparison of survival curves for different combinations of risk factors, and uses statistical models to facilitate this comparison.

Methods for analysing data from follow-up studies are described in Chapter 6, and these methods also provide the basic tools for analysing survival data. If the risk of an event is small, Poisson regression may be used to model incidence and to estimate relative risks. Even if these risks are not small logistic regression may be applied to survival data, as you saw in the comparison of patients' departure rates from a psychiatric ward considered in Section 6 of the preceding chapter.

The methods of Chapter 6 are quite appropriate for the analysis of data grouped by duration of survival, such as typical data arising in actuarial science and demography. But if the survival times are measured on a continuous scale, more straightforward and general methods, using the case-by-case data structure, are available. These methods are described in the present chapter.

In survival analysis the outcome variable of interest is the time until an event occurs. This duration could be measured in days, weeks, months or years from the beginning of observed follow-up on a subject. As usual the event could be death, disease incidence or relapse, recovery or partial recovery, or generally any designated occurrence to an individual.

In general, survival analysis allows for the proper treatment of incomplete data due to subjects dropping both into and out of the study. However, the methods described in this chapter will deal only with drop-outs, giving rise to *censored* (more precisely, *right-censored*) data. In fact survival data may be censored for any of the following three reasons.

(a) a subject withdraws from the study for any reason before experiencing the event (this includes so-called 'loss to follow-up');

(b) an intervening event occurs (such as a failure from an unrelated cause), prohibiting further observation on that subject;

(c) a subject does not experience the event before the study ends (or an analysis of the results is required).

Figure 7.1 graphically illustrates a situation involving a small number of subjects entering a follow-up study. Each subject's observed time in the study is represented by a horizontal line. Note that in practice the subjects do not enter the study at the same time: there is an *accrual period* during which subjects are recruited (weeks 1 to 16 for the data graphed), then a further follow-up period (e.g., weeks 17 to 36), at the end of which an analysis is needed. Thus, subjects who have still not experienced the event at the end of the follow-up period provide censored data of type (c). Incomplete data of types (a) and (b) correspond to subjects whose survival times are truncated during the accrual or follow-up period due to withdrawals or intervening events.

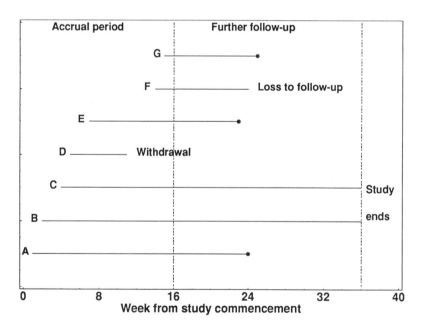

FIGURE 7.1: Accrual and Survival in a Follow–up Study

For the data shown in Figure 7.1, subjects A, E and G enter in Weeks 1, 7 and 15, respectively, experiencing the event (represented by a filled circle) in Weeks 24, 23 and 25; B and C enter in Weeks 2 and 4, respectively, and neither has experienced the event by Week 36, when the follow-up period ends; D enters in Week 5 and withdraws in Week 11, while F enters in Week 14 and is lost due to accidental death (say) in Week 24.

Survival times after which the event of interest is observed to occur are conventionally called *failure times* (even if the event is a 'success' such as recovery from some disease).

In the next section you will learn how to construct a survival curve from data such as those shown in Figure 7.1.

2: The Survival Curve

The survival curve is defined as the proportion of subjects surviving beyond a given duration of time t. For a large population, this curve will be a smooth function of t which decreases from a maximum value 1 when t is 0. In practice the survival curve estimated from a (necessarily finite) sample of data is a step function which only decreases at the times when events occur.

The graphical construction of a survival curve based on some data is reasonably straightforward, and is not much more complicated than plotting the empirical distribution of an ordinary sample of numerical data. To illustrate the method of construction, consider the data graphed in Figure 7.1.

First imagine that all subjects had entered the study at the same time origin: this involves shifting the line segments representing their observed survival times so that they all start at $t = 0$, as shown in Figure 7.2. The *number at risk* at time t is then defined as the number of observed survival times exceeding a duration t. In our small data set, for example, the number at risk is 7 when t is 6 weeks or less, 6 for t in the range 7 to 10 weeks, after which time it drops to 4, since two survival times end (one censored and one complete) at Week 10. Beyond Week 10, the number at risk drops by one further unit at weeks 16, 23 and 32, reaching 0 after Week 34.

The final step in the construction of a survival curve involves setting up a table with each row corresponding to a failure time. The table should contain columns containing (a) the failure time (t), (b) the number of failures at each such time (F_t), (c) the number of subjects remaining at risk *just before* that time (R_t), and (d) the proportion of survivors ($1 - F_t / R_t$). Table 7.1 shows this information for the data graphed in Figures 7.1 and 7.2. In this case there are only three failure times, at weeks 10, 16 and 23 after entry of a subject, so the table has just three rows. Note that there are two survival times both equal to 10 weeks, one censored because of loss to follow-up, the other a failure time. To compute the number at risk at such times, the censored data are included in the number at risk, so the number at risk just before 10 weeks is 6. Of course if the subject had been lost to follow-up just a little bit sooner, the number at risk corresponding to 10 weeks would be 5 instead of 6.

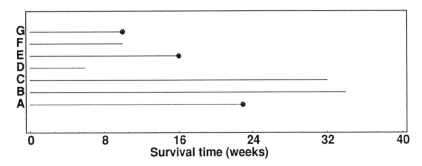

FIGURE 7.2: The Number of Subjects at Risk (for Data from Figure 7.1)

t	F_t	R_t	$1 - F_t/R_t$	S_t
10	1	6	0.833	0.833
16	1	4	0.75	0.625
23	1	3	0.667	0.416

TABLE 7.1: Calculation of Survival Curve

The estimate of the survival curve at each failure time t (S_t) is now obtained by multiplying together the proportions of survivors up to and including the given failure time. The curve remains horizontal between each failure time. So for the data given in Table 7.1, the survival curve is 1 up to 10 weeks, at which point it drops to 5/6 = 0.833. It then remains at 0.833 until Week 16, where it drops to 0.833 × (3/4) = 0.625, remaining at this value until Week 23, where it decreases further to 0.625 × (2/3) = 0.416, remaining at this value until Week 34, where it is truncated because of lack of further information.

The estimated survival curve obtained in this way is called the *Kaplan-Meier* curve (see Kaplan and Meier (1958)), and is graphed in Figure 7.3 for the data from Figure 7.1.

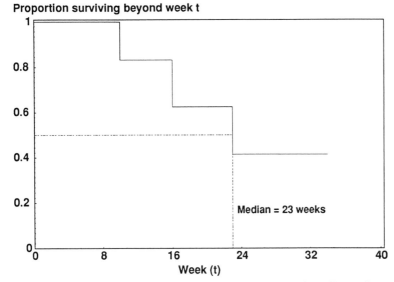

FIGURE 7.3: Kaplan–Meier Survival Curve for Subjects from Figure 7.1

A useful summary of survival that may be obtained directly from a survival curve is the *median survival duration*. This is the survival time exceeded by 50% of the subjects, and is obtained simply by determining where the value of the survival curve is 0.5. Graphically, you just draw a horizontal line cutting the vertical axis at 0.5, find where it intersects the survival curve, and then drop a perpendicular line from that point to the time axis. For the survival curve graphed as Figure 7.3, the median survival time is thus 23 weeks. It may happen (due to excessive censoring) that the survival curve never falls below the 0.5 line. In such cases the median is not computable: all you can say is that the median survival time is greater than the largest observed failure time.

While the survival curve for a group of subjects participating in a follow-up study is of interest in its own right, it is usually of greater interest to *compare* the survival curves for different groups of subjects. To do this, you simply plot the survival curves for the different groups on the same axes. The higher the curve, the better the survival prospects for the subjects. Thus if one curve (for Group A, say) is completely above another (Group B), then the subjects in Group A have better survival prospects than those in Group B. However, if the two curves cross, the situation is more complicated: it means that the relative risk of failure depends on how long a subject has survived. In the next section the problem of how to compare survival curves is considered.

3: Comparing Survival Curves

The small sample of survival times for the seven subjects shown in Figure 7.1 is actually a subset of the remission times of 42 leukemia patients studied by Freireich et al (1963), divided into a treatment group and a control group. The survival time and corresponding *status* (censored or complete) for each patient is given in Table 7.2.

Treatment Group	Control Group
6, 6, 6, 7, 10,	1, 1, 2, 2, 3,
13, 16, 22, 23,	4, 4, 5, 5,
6+, 9+, 10+, 11+,	8, 8, 8, 8,
17+, 19+, 20+,	11, 11, 12, 12
25+, 32+, 32+,	15, 17, 22, 23
34+, 35+	

+ denotes censored data

TABLE 7.2: Remission Times (weeks) of Leukemia Patients

There are 21 subjects in each group. Since none of the survival times in the control group is censored, the survival curve for these patients is very easy to construct: it just drops by the amount 1/21 at each failure time. If two or more failure times are equal (or *tied*), the amount by which the estimated survival curve drops at this point is equal to the sum of the individual components. Thus the survival curve for the control group drops from 1 to $1-(2/21)$ at $t = 1$ week, and by the same amount at weeks 2, 4, 5, 11 and 12: at each of these times there are two tied failure times. It drops by the amount 4/21 at Week 8, where there are four ties. Check this for yourself by constructing a table analogous to Table 7.1 (containing 21 rows corresponding to the 21 failure times in this group).

You may develop a table, similar to Table 7.1 but for the larger sample, and consequently construct the Kaplan–Meier curve for the patients in the treated group. The survival curves for the two treatment groups are graphed together in Figure 7.4. You can see from this that the patients in the treated group remain in remission longer than those in the control group, since the survival curve for the treated patients is higher. The median survival times are also shown: 23 weeks for the treated group compared with only 8 weeks for the control group.

The curves certainly *look* different, but a p-value is needed for confirmation. A basis for comparing the curves is to compute an odds ratio or relative risk. These estimates may be adjusted for the possible confounding effect of survival duration using the Mantel–Haenszel methods outlined in Chapter 4.

Proportion surviving beyond week t

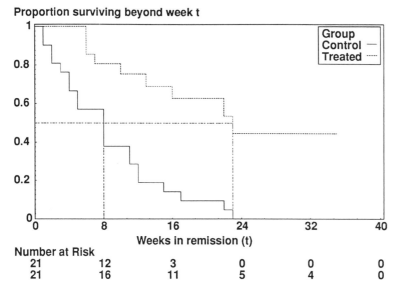

FIGURE 7.4: Kaplan–Meier Curves for two Groups of Remission Times

The average risk of relapse to a patient in any given week is small (30 relapses in 541 person-weeks), so the methods for analysing incidence density ratios given in Section 4 of Chapter 4 may be used. There are $a = 9$ relapses and $L_1 = 359$ person-weeks total exposure time in the treatment group, compared with $b = 21$ events in the control group with $L_0 = 182$ person-weeks exposure, so Equation (4.11) gives $(9/359) / (21/182) = 0.217$ for the estimated incidence density ratio. Using Equation (4.12), the standard error of its logarithm is $\sqrt{(1/9+1/21)} = 0.398$, giving a 95% confidence interval $(0.100, 0.474)$. A z-statistic is obtained by dividing the logarithm of the estimate by its standard error: this gives $z = -1.527/0.398 = -3.83$, so the p-value is 0.0001, strong evidence that the relative risk is below 1. This assumes that the risks of relapse do not depend on the elapsed time.

Now consider the analysis using duration of remission as a stratification variable. The stratified data are shown in Table 7.3. Note that this table has a similar layout to Table 6.29.

Having stratified the data by failure time, alternative methods are available for comparing the survival curves. The Mantel–Haenszel methods given in Chapter 4 give adjusted estimates of the relative risks. P-values may be obtained from chi-squared tests based on comparing observed and expected numbers of events at the failure times. Finally (as you saw in Section 6 of Chapter 6) logistic (or Poisson) regression may be used to fit a model, providing estimates of relative risks as well as p-values. These methods are described next.

| Treatment Group | | Control Group | | Stratum |
Failures	Exposed	Failures	Exposed	(Failure time)
0	21	2	21	1
0	21	2	19	2
0	21	1	17	3
0	21	2	16	4
0	21	2	14	5
3	21	0	12	6
1	17	0	12	7
0	16	4	12	8
1	15	0	8	10
0	13	2	8	11
0	12	2	6	12
1	12	0	4	13
0	11	1	4	15
1	11	0	3	16
0	10	1	3	17
1	7	1	2	22
1	6	1	1	23

TABLE 7.3: Data Layout – Leukemia Remissions Stratified by Failure Time

Mantel–Haenszel Adjustment

For the data in Table 7.3 Equation (4.6) gives an estimate 0.192 for the Mantel–Haenzsel adjusted odds ratio, with a 95% confidence interval (using the Robins formula given by Equation (4.7)) 0.080 to 0.456. The crude estimate, based on aggregating the data and using Equation (4.1), is $(9/247) / (21/141) = 0.245$, indicating some confounding with respect to failure time.

Note that these estimates, in contrast to the crude estimate of the incidence density ratio given earlier in this section, do not take the duration of survival into account. They also assume that the events are completely independent, and are based on the 418 separate opportunities for a relapse to occur, rather than on just 42 subjects. Essentially it is being assumed that the subject 'forgets' what has happened in the past, giving independent risks of relapse at the various times.

The Logrank Test

The Mantel–Haenszel test for comparing survival curves is more commonly known as the *logrank* test. It is similar to a chi-squared test for comparing two proportions, but is more complex, in the sense that it is based on a sum of components, where each component corresponds to a different failure time. To see how the statistic is defined, first consider a particular failure time at which the observed numbers of failures occurring in the two groups are f_1 and f_2 respectively, and the corresponding numbers at risk are r_1 and r_2. Then under the null hypothesis that the population survival curves are the same, the *expected allocation* of failures to the two groups (given that $f_1 + f_2$ failures have occurred) is in proportion to the numbers at risk in the two groups, and is consequently given by $e_1 = (f_1 + f_2)r_1/(r_1 + r_2)$ for the first group and $e_2 = (f_1 + f_2)r_2/(r_1 + r_2)$ for the second group.

To illustrate these calculations, look at $t = 8$ weeks for the data listed in Table 7.3. Four failures occurred, all in the second (control) group, so f_1 is 0 and f_2 is 4. The numbers at risk are $r_1 = 16$ and $r_2 = 12$, so the expected allocation is $e_1 = 4 \times (16/28) = 2.29$ and $e_2 = 4 \times (12/28) = 1.71$. Now look at $t = 22$ weeks, where there were two failures, one in each group, so f_1 and f_2 are both 1. The numbers at risk have now dropped to $r_1 = 7$ and $r_2 = 2$, giving an expected allocation of $1 \times (7/9) = 0.778$ and $1 \times (2/9) = 0.222$.

Adding up all these contributions at the different failure times, you get $E_1 = \Sigma e_1$ and $E_2 = \Sigma e_2$ for the total *expected* numbers of failures in the two groups. These may then be compared with the total *observed* numbers of failures $O_1 = \Sigma f_1$ and $O_2 = \Sigma f_2$ for the two groups, using a chi-squared test to give a p-value. The number of degrees of freedom in this test is just 1, the same as in a simple chi-squared test for comparing two population proportions.

In this case the values of E_1 and E_2 turn out to be 19.25 and 10.75, respectively, whereas O_1 is 9 and O_2 is 21. An approximate chi-squared statistic is obtained in the usual way by comparing the observed and expected counts using the formula $\Sigma(O_i - E_i)^2/E_i$, which gives 15.23. However, there is a better approximation to a chi-squared statistic based on the covariances between the numbers of failures (see Kalbfleisch and Prentice (1980), pages 79–81), which gives the result 16.79 for the logrank test statistic in the present case. Comparing 16.79 with a chi-squared distribution with 1 degree of freedom, the p-value is less than 0.0001. This result is identical to that obtained from Equation (4.8) for the Mantel–Haenszel test.

The logrank test attaches equal weight to all failure times. Another test proposed by Richard Peto (see Prentice and Marek (1979)), which also has a chi-squared distribution with 1 degree of freedom, attaches greater weight to early failures (those closer to $t = 0$). In the present example, its value turns out to be 9.95 (p-value 0.0016).

Both the logrank test and Peto's modification extend in a natural way to situations where there are more than two survival curves to be compared. The null hypothesis states that all the populations generating the survival times are the same. The only difference is that the test statistics in the more general case need to be compared with a chi-squared distribution with $c - 1$ degrees of freedom, where c is the number of curves being compared.

Poisson Regression

The upper printout in Table 7.4 shows what happens when a Poisson regression model is fitted to the data in Table 7.3, taking subjects with failure times equal to 1 week as the referent group. As you can see from this printout, the estimate of the incidence density ratio given by this method is 0.221 with a 95% confidence interval (0.10, 0.49), consistent with the result given by the Mantel–Haenszel method.

```
Poisson Regression Analyses
Response: Relapse from Remission
  Col Name         Coeff  StErr p-value     IDR    95%      CI
```

Col	Name	Coeff	StErr	p-value	IDR	95%	CI
0	Constant	-2.551	0.709	0.000			
3	Treatment	-1.509	0.409	0.000	0.221	0.099	0.493
6	Week=2	0.081	0.998	0.935	1.085	0.153	7.671
7	Week=3	-0.524	1.223	0.669	0.592	0.054	6.514
8	Week=4	0.217	0.998	0.828	1.242	0.176	8.788
9	Week=5	0.319	0.998	0.750	1.375	0.194	9.735
10	Week=6	0.838	0.912	0.358	2.311	0.387	13.799
11	Week=7	-0.206	1.223	0.866	0.814	0.074	8.948
12	Week=8	1.194	0.864	0.167	3.301	0.607	17.952
13	Week=10	0.125	1.224	0.919	1.133	0.103	12.477
14	Week=11	0.858	0.999	0.390	2.358	0.333	16.697
15	Week=12	1.086	0.999	0.277	2.963	0.418	21.013
16	Week=13	0.656	1.226	0.593	1.927	0.174	21.321
17	Week=15	0.690	1.226	0.574	1.993	0.180	22.029
18	Week=16	0.859	1.228	0.484	2.360	0.213	26.197
19	Week=17	0.900	1.227	0.463	2.461	0.222	27.265
20	Week=22	1.978	1.004	0.049	7.228	1.011	51.671
21	Week=23	2.400	1.011	0.018	11.022	1.520	79.908

```
df:16              Dev: 27.634       #iter:9        RSq:0.477
```

Col	Name	Coeff	StErr	p-value	IDR	95%	CI
0	Constant	-2.043	0.218	0.000			
3	Treatment	-1.305	0.397	0.001	0.271	0.125	0.59

```
df:32              Dev:  40.81       #iter:7        RSq:0.227
```

TABLE 7.4: Alternative Poisson Models for Leukemia Remissions

The lower printout in Table 7.4 shows the effect of omitting failure time as a covariate in the model. The deviance has increased by 13.18 (from 27.63 to 40.81) but this is less than its expected value, the number of additional degrees of freedom (16), so there appears to be no benefit in including failure time as a covariate in the model.

Note that the simple Poisson model gives an estimate of relative risk 0.271 which is higher than the value 0.217 corresponding to the crude incidence density ratio estimate based on Equation (4.11). This discrepancy arises because the crude estimate takes duration of survival into account (including the intervals between failures), whereas the Poisson estimate is based on a simple aggregation of the data in Table 7.3 (ignoring the intervals between failures).

The conclusion from the analyses is that there is a difference in survival between the treatment and control groups, and the relative risk of relapse does not depend on the duration of remission.

As a further illustration, Table 7.5 shows data from a random-ised control clinical trial investigating prednisolone therapy reported by Kirk et al (1980) and discussed in Pocock (1986, Table 14.6). These are survival times in months until death for chronic active hepatitis patients. Kaplan–Meier curves are shown in Figure 7.5.

Treatment Group	Control Group
2, 6, 12, 54, 56+,	2, 3, 4, 7, 10,
68, 89, 96, 96, 125+	22, 28, 29, 32,
128+, 131+, 140+,	37, 40, 41, 54,
141+, 143, 145+,	61, 63, 71, 127+,
146, 148+, 162+,	140+, 146+, 158+,
168, 173+, 181+	167+, 182+

+ denotes censored data

TABLE 7.5: Survival Times (months) of Hepatitis Patients

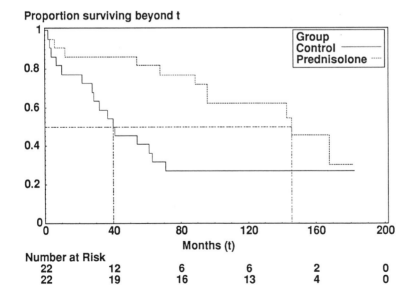

FIGURE 7.5: Kaplan–Meier Curves for Hepatitis Patients

Table 7.6 shows the data layout for analysis with failure time as a stratification variable. The logrank test (*p*-value 0.031) and Peto's test (*p* = 0.034) give similar results, indicating a statistically significant treatment benefit. The Mantel–Haenszel adjusted relative risk is 0.42 with 95% confidence interval (0.19, 0.94).

| Treatment Group | | Control Group | | Stratum |
Failures	Exposed	Failures	Exposed	(Failure time)
1	22	1	22	2
0	21	1	21	3
0	21	1	20	4
1	21	0	19	6
0	20	1	19	7
1	20	1	18	10
0	19	0	17	12
0	19	1	17	22
0	19	1	16	28
0	19	1	15	29
0	19	1	14	32
0	19	1	13	37
0	19	1	12	40
0	19	1	11	41
1	19	1	10	54
0	17	1	9	61
0	17	1	8	63
1	17	0	7	68
0	16	1	7	71
1	16	0	6	89
2	15	0	6	96
1	8	0	4	143
1	6	0	4	146
1	3	0	1	168

TABLE 7.6: Data Layout for Hepatitis Survival Stratified by Failure Time

Table 7.7 gives printouts from various Poisson regression analyses. The first two printouts match those in Table 7.4, and show that omitting failure time as a covariate increases the deviance by only 7.96 (from 27.97 to 35.93). Since failure time has 22 degrees of freedom, this deviance increase is less than its expected value and is thus not statistically significant. However, the coefficients associated with the failure times follow a trend which is approximately linear. Including failure time as a continuous variable (expressed as the number of months elapsed) gives the third printout in Table 7.7, which has the following simple interpretation.

For a patient in the target population, prednisolone treatment is associated with a reduced risk of death at any time (estimated relative risk 0.44, 95% confidence interval 0.20–0.95). Irrespective of treatment, the risk increases with time by a factor estimated to be 1.012 (95% confidence interval 1.003–1.020) for each month of survival.

```
Poisson Regression Analyses
Response: Died
Col Name          Coeff StErr p-value   IDR   95%      CI

 0 Constant      -2.759 0.710  0.000
 3 Prednisolone -0.832 0.397   0.036 0.435 0.200   0.947
 6 Month=3      -0.647 1.220   0.596 0.524 0.048   5.720
 7 Month=4      -0.613 1.220   0.615 0.542 0.050   5.916
 8 Month=6      -0.578 1.220   0.636 0.561 0.051   6.127
 9 Month=7      -0.562 1.220   0.645 0.570 0.052   6.223
10 Month=10     -0.526 1.220   0.667 0.591 0.054   6.456
11 Month=12     -0.487 1.220   0.689 0.614 0.056   6.708
12 Month=22     -0.470 1.220   0.700 0.625 0.057   6.823
13 Month=28     -0.430 1.220   0.724 0.651 0.060   7.105
14 Month=29     -0.388 1.220   0.751 0.679 0.062   7.411
15 Month=32     -0.344 1.220   0.778 0.709 0.065   7.746
16 Month=37     -0.298 1.220   0.807 0.742 0.068   8.113
17 Month=40     -0.250 1.220   0.838 0.779 0.071   8.518
18 Month=41     -0.199 1.221   0.870 0.819 0.075   8.965
19 Month=54      0.547 0.996   0.583 1.729 0.245  12.171
20 Month=61     -0.038 1.221   0.975 0.963 0.088  10.543
21 Month=63      0.025 1.222   0.984 1.025 0.094  11.241
22 Month=68      0.092 1.223   0.940 1.097 0.100  12.042
23 Month=71      0.123 1.222   0.920 1.131 0.103  12.408
24 Month=89      0.197 1.223   0.872 1.218 0.111  13.393
25 Month=96      0.925 0.998   0.354 2.521 0.357  17.819
26 Month=143     0.747 1.221   0.541 2.110 0.193  23.122
27 Month=146     0.871 1.220   0.476 2.388 0.218  26.105
28 Month=168     1.924 1.224   0.116 6.848 0.622  75.444

 df:23       Dev: 27.972       #iter:8       RSq:0.287

Col Name          Coeff StErr p-value   IDR   95%      CI

 0 Constant      -2.918 0.249  0.00
 3 Prednisolone -0.705 0.390   0.07 0.494     0.23 1.061

 df:46       Dev: 35.931       #iter:7       RSq:0.085

Col Name          Coeff StErr p-value   IDR   95%      CI

 0 Constant      -3.351 0.319  0.000
 3 Prednisolone -0.822 0.393   0.037 0.440 0.204   0.95
 4 Month         0.012 0.004   0.007 1.012 1.003   1.02

 df:45       Dev: 29.955       #iter:7       RSq:0.237
```

TABLE 7.7: Alternative Poisson Models for Survival of Hepatitis Patients

In this section we have described three methods for comparing survival curves: Mantel–Haenszel adjustment of the relative risk after stratifying by time to failure, testing the null hypothesis that the population survival curves are identical, and Poisson regression again using failure time as a stratification variable. The first two methods are satisfactory if there are no other factors (covariates or explanatory variables) which may affect survival. Poisson regression handles additional covariates, but the data layout is complicated, particularly if the number of different failure times is large. We give a simpler method in the next section.

4: The Proportional Hazards Model

Recall that the Poisson regression model (6.1) specifies the logarithm of the incidence density at any time t as a linear function of covariates. This incidence density is actually the risk *rate*, that is, the probability that the outcome event of interest occurs in a small interval $(t, t + dt)$ divided by the length of the interval (dt). The model assumes that the incidence density does not depend on the elapsed time t.

In survival analysis we want to generalise this model to allow the incidence density to depend on t, and the term *hazard function* is used instead of incidence density. Assume that this model is multiplicative, so that the incidence density is the product of a function of t and an exponential linear function of the covariates, that is,

$$h_t = h_{0t} \exp(\sum_{j=1}^{p} b_j x_j) \qquad (7.1)$$

where h_{0t} is the *baseline* hazard function, that is, the hazard function corresponding to an individual or group for which the value of each covariate is 0. Note that the constant term a occurring in Equation (6.1) is omitted from Equation (7.1), because it may be absorbed into the baseline hazard function.

In fact Equations (6.1) and (7.1) are more similar than they appear at first sight. Suppose that the covariates x_j in Equation (6.1) include a stratification variable for failure time, as for the examples given in the preceding section. For m failure times $t_0, t_1, ..., t_{m-1}$, this stratification variable may be coded as $m - 1$ indicator variables, denoted by $w_1, w_2, ..., w_{m-1}$, say. Equation (7.1) then takes the form

$$\lambda = \exp(a + \sum_{j=1}^{q} b_j x_j + \sum_{j=q+1}^{m+q} b_j w_j) \qquad (7.2)$$

where q is the number of covariates apart from the stratification variable. In this formulation, each w_j is an indicator variable taking the value 1 for a failure occurring at time t_j and 0 otherwise.

Equation (7.2) may be rewritten as

$$\lambda = \exp(a + \sum_{j=q+1}^{m+q} b_j w_j) \times \exp(\sum_{j=1}^{q} b_j x_j) \qquad (7.3)$$

Now the first factor in the right-hand side of Equation (7.3) contains only the stratification variable, which is a function of the failure times, but does not involve the other covariates. It follows that Equations (7.1) and (7.3) are equivalent.

Equation (7.1) is called the *proportional hazards model*. It was considered as a model for survival data by Cox (1972). The term 'proportional hazards' arises from the fact that the relative risk of failure or *hazard ratio* at time t for any two subjects i and k is given by the ratio of their hazard functions. Using the model, this relative risk does not depend on t because the common factor h_{0t} cancels, so that

$$\frac{h_{it}}{h_{kt}} = \exp\left(\sum_{j=1}^{p} b_j(x_{ij} - x_{kj})\right)$$

Consequently the hazards are always in constant proportion: their ratio does not vary with the survival duration.

The data layout needed to fit the model (7.1) to survival data is much simpler than that needed for Poisson regression. The data are stored in the case-by-case format with one column containing the survival times, another containing the survival status (complete or censored), and the remaining columns containing the values of the covariates. To illustrate, consider the remission times of the leukemia patients given in Table 7.2, for which the data comprise 42 records in three columns as shown in Table 7.8 below.

Note that the censoring status s is coded as 0 (censored) or 1 (completed event), and the covariate x is coded as 1 (treated patient) or 0 (control patient).

The model (7.1) may be fitted to data by maximising a likelihood associated with the configuration of survival times, a procedure that produces the printout in Table 7.9 for the data in Table 7.8. This result may be compared with the first printout given in Table 7.4, where Poisson regression is used with failure time as a stratification variable. The estimates and standard errors for the treatment effect are identical (allowing for small roundoff errors in the computations).

t	s	x		t	s	x
6	1	1		1	1	0
6	1	1		1	1	0
6	1	1		2	1	0
7	1	1		2	1	0
10	1	1		3	1	0
13	1	1		4	1	0
16	1	1		4	1	0
22	1	1		5	1	0
23	1	1		5	1	0
6	0	1		8	1	0
9	0	1		8	1	0
10	0	1		8	1	0
11	0	1		8	1	0
17	0	1		11	1	0
19	0	1		11	1	0
20	0	1		12	1	0
25	0	1		12	1	0
32	0	1		15	1	0
32	0	1		17	1	0
34	0	1		22	1	0
35	0	1		23	1	0

TABLE 7.8: Simplified Data Layout for Modelling Leukemia Remission Times

```
Cox Regression Analysis
Response: Remission (weeks)
 Col Name      Coeff StErr p-val HazRat     95%    CI P(PH)
```

	Col Name	Coeff	StErr	p-val	HazRat	95%	CI	P(PH)
3	Treated	-1.509	0.410	0	0.221	0.099	0.493	0.794

```
 n:42         %Cen: 28.571      -2LogL: 172.759      #iter:7
```

TABLE 7.9: Proportional Hazards Model: Leukemia Remission Times

The method also gives estimated survival curves based on the model (see Kalbfleisch and Prentice (1980), pages 84–87), and these curves are shown in Figure 7.6. Note that they are very similar to the Kaplan–Meier curves for the same data shown in Figure 7.4. The survival curves from the model are also smoother than the Kaplan–Meier curves.

Judging by the closeness of the proportional hazards (PH) model survival curves to the Kaplan–Meier curves, the PH model appears to provide a good fit to these data. But now look at Table 7.10 and Figure 7.7, which show the printout and survival curves after fitting the model to the data for the hepatitis patients listed in Table 7.5.

Proportion surviving beyond week t

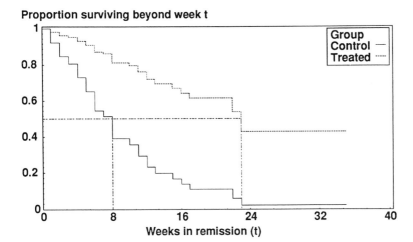

FIGURE 7.6: Proportional Hazards Survival Curves: Leukemia Patients

```
Cox Regression Analysis
Response: Survival (months)
  Col Name      Coeff StErr p-val HazRat    95%     CI P(PH)

   3 prednis.  -0.832 0.397 0.036  0.435   0.2 0.948 0.153

 n:44          %Cen: 38.636    -2LogL: 173.436      #iter:7
```

TABLE 7.10: PH Model fitted to Hepatitis Patients' Survival TIMES

Proportion surviving beyond t

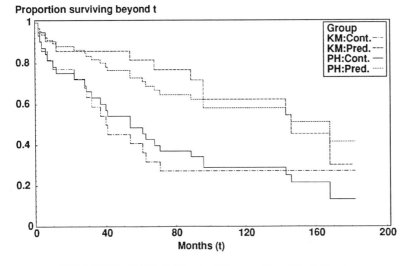

FIGURE 7.7: PH Model Survival Curves: Hepatitis Patients

To facilitate the comparison both sets of curves are shown in Figure 7.7, and you can see that the Kaplan–Meier survival curves are wider apart than the model-based curves.

The printouts shown in Tables 7.9 and 7.10 give p-values for assessing the validity of the proportional hazards assumption in the column labelled P(PH): small p-values indicate that the assumption is violated. This test is due to Schoenfeld (1982) and Harrell and Lee (1986). For the leukemia remission times this p-value is 0.79, indicating that the fit is quite acceptable, whereas the corresponding p-value for the hepatitis survival times is 0.15, suggesting that the fit of the PH model is not as good (though still acceptable), and confirming the pattern in Figure 7.7.

The PH model expresses the instantaneous risk of failure as a function of the risk factors of interest, whereas the survival curve estimates the probability of surviving up to a specified time. In the next section we examine the relation between the hazard function and the survival curve.

5: Hazard versus Survival

It may be shown that for a given survival curve S_t, the corresponding hazard function is given by the derivative of $-\ln S_t$. Equivalently the survival curve may be expressed as $\exp(-H_t)$, where H_t is the integrated hazard function.

To get some feeling for the relation between hazard functions and survival curves, first imagine that the hazard function is constant. In other words, assume that the risk of failure is not affected by how long the subject has been observed: it neither increases nor decreases with time. The top panel of Figure 7.8 shows a family of such constant hazard functions with respective failure risk rates of 1, 2 and 3 per unit time.

Using the mathematical relation between the hazard h_t and the proportion surviving S_t, the survival curve corresponding to a hazard function taking the constant value λ may be shown to follow the exponential distribution

$$S_t = e^{-\lambda t} \tag{7.4}$$

The lower panel of Figure 7.8 shows these survival curves, corresponding to the hazard functions in the top panel. Note that the higher the hazard, the more rapidly the survival curve decreases with time.

Similarly, Figures 7.9 and 7.10 show non-constant hazard functions with their corresponding survival curves.

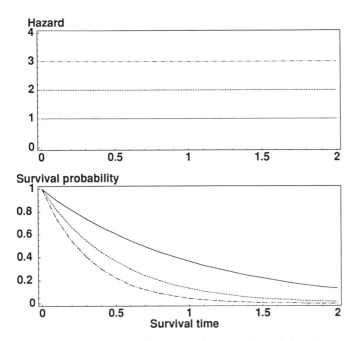

FIGURE 7.8: Survival Curves for Constant Hazard Functions

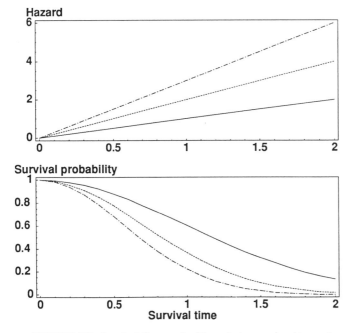

FIGURE 7.9: Survival Curves for Linearly Increasing Hazards

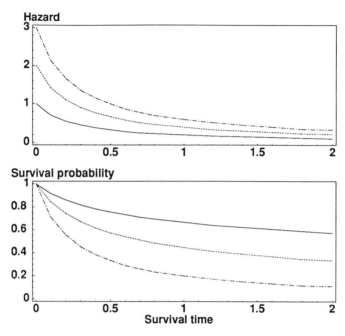

FIGURE 7.10: Survival Curves for Decreasing Hazards

The hazard functions in Figure 7.9 differ from each other by simple factors of 2 and 3 and increase linearly with time. In this case you can see that the survival curves eventually drop quite sharply to 0. The hazard functions in Figure 7.10 take the form $h(t) = c/(1 + 4t)$, with the constant c taking the values 1, 2 and 3.

The survival curve summarises the cumulative risk of failure, and its value at time t is the probability that a subject will not fail in the interval $(0, t)$. On the other hand the hazard function describes the risk of failure at a given time point t and nowhere else.

The proportional hazards model expresses the hazard function as a product of a baseline hazard function and a constant of proportionality which depends on the values of the predictor variables: the formula is given as Equation (7.1). Using the relation between a hazard function and its corresponding survival curve, the model may be written as

$$S_{it} = S_{0t}^{\exp(\Sigma b_j x_{ij})} \qquad (7.5)$$

where S_{0t} is the *baseline survival curve*.

The baseline survival curve may be estimated from the data after fitting a proportional hazards model. It is then possible to obtain a model survival curve for any specified values of the covariates, simply by substituting appropriate values into the above formula. This is how the model curves graphed in Figures 7.6 and 7.7 are obtained.

Graphical Assessment of the PH Assumption

Equation (7.5) provides the basis for a graphical method for assessing the proportional hazards assumption. You can get some feel for the validity of this assumption when comparing two or more survival curves by transforming the survival curves, by taking logarithms twice, before plotting them. The rationale for this transformation can be seen by examining the formula for the survival curve in the PH model. Taking natural logarithms in Equation (7.5) and then reversing signs, you get

$$-\ln(S_{it}) = -R_i \ln(S_{0t}) \qquad (7.6)$$

Since S_t is necessarily between 0 and 1, its logarithm is negative, so reversing the sign makes each side of the above equation positive. Taking natural logarithms and reversing signs again, you get

$$-\ln(-\ln(S_{it})) = -\ln(-\ln(S_{0t})) - \ln(R_i) \qquad (7.7)$$

Now consider the curves corresponding to two groups in the population, indexed by i and j. The difference between their transformed curves is just $\ln(R_i) - \ln(R_j)$, which is constant, not depending on t. So if you plot these transformed curves for the two subgroups on the same axes, they should be parallel if the PH assumption is met.

Figure 7.11 shows the transformed Kaplan–Meier survival curves for the two examples considered earlier. For the leukemia remission times the curves are approximately parallel so the proportional hazards assumption is reasonable. However, the curves for the hepatitis patients tend to diverge at first and then converge, so the proportional hazards assumption is questionable. These graphical patterns are consistent with the p-values for the PH assumption given in the printouts in Tables 7.9 and 7.10.

The two examples considered so far involve just one covariate or risk factor. The proportional hazards model, like logistic and Poisson regression, allows for any number of covariates. Examples involving several covariates are considered in the next section.

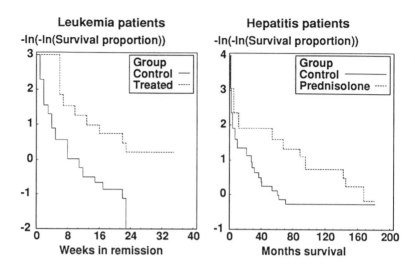

FIGURE 7.11: Diagnostic Curves for Leukemia and Hepatitis Patients

6: Modelling Covariates

Leukemia is a cancer characterised by an over proliferation of white blood cells: the higher this count, the greater the severity of disease. The data listed below, cited by Feigl and Zelen (1965), are survival times (in weeks) and white blood counts (WBC) for two groups of patients who died of acute myelogenous leukemia, classified according to presence or absence of a morphologic characteristic of white cells. Those termed AG positive had Auer rods and/or significant granulature of the leukemic cells in the bone marrow at diagnosis.

The white blood cell counts have a very skewed distribution, and are better expressed as logarithms. The printout in Table 7.12 shows what happens when the proportional hazards model is fitted to these data. The data layout is as in Table 7.8, with status s equal to 1 for each case (given that none of the survival times is censored) and covariates x_1 (coded as 1 for the positive AG group and 0 for the negative AG group) and x_2 (base 10 logarithm of white blood cell count).

The printout shows no evidence of a departure from the proportional hazards assumption, since the associated p-values are not small. The effect of AG positivity is to reduce the risk of death at any time by a factor 0.36 (95% CI: 0.16–0.83) while a tenfold increase in white blood count increases the risk by 2.29 (95% CI: 1.24–4.23).

AG positive		AG negative	
WBC	Time to death	WBC	Time to death
2300	65	4400	56
750	156	3000	65
4300	100	4000	17
2600	134	1500	7
6000	16	9000	16
10500	108	5300	22
10000	121	10000	3
17000	4	19000	4
5400	39	27000	2
7000	143	28000	3
9400	56	31000	8
32000	26	26000	4
35000	22	21000	3
100000	1	79000	30
100000	1	100000	4
52000	5	100000	43
100000	65		

TABLE 7.11: Survival Times of Leukemia Patients

```
Cox Regression Analysis
Response: Survival (weeks)
 Col Name        Coeff StErr p-val HazRat     95%    CI P(PH)

   3 AG+        -1.018 0.423 0.016  0.361  0.158 0.829 0.472
   4 log(WBC)   0.830 0.312 0.008  2.292  1.244 4.226 0.857

 n:33          %Cen: 0.000      -2LogL: 157.363      #iter:7
```
TABLE 7.12: PH Model fitted to Leukemia Survival Data

Table 7.13 shows the effect of omitting white blood count from the model. The estimated hazard ratio for AG positivity has changed only slightly, from 0.36 to 0.33, so the white blood count is not a significant confounder.

```
Cox Regression Analysis
Response: Survival (weeks)
 Col Name        Coeff StErr p-val HazRat     95%    CI P(PH)

   3 AG+        -1.116 0.412 0.007  0.328  0.146 0.735 0.560

 n:33          %Cen: 0.000      -2LogL: 164.466      #iter:7
```
TABLE 7.13: PH Model with WBC omitted: Leukemia Survival Data

Model-based survival curves may be constructed for any combination of values of the covariates, using Equation (7.5), and are thus expressible as powers of a baseline survival curve. Figure 7.12 shows the curves corresponding to the four groups defined by AG+ and AG− and the lower and upper quartiles of the white blood counts.

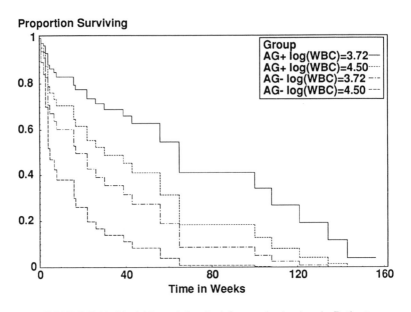

FIGURE 7.12: Model-based Survival Curves for Leukemia Patients

It is also possible to construct survival curves for specified covariate values after adjusting for the effect of other covariates. For example, you may wish to compare survival curves for the two AG groups after adjusting in some way for the effect of the other covariate, x_2. One way of doing this is to replace x_2 by a typical value such as its mean or median, and to plot the model-based survival curves for the two treatment groups with this value of x_2. For the present example these curves are given by the formula

$$\bar{S}_{it} = S_{0t}^{\exp(b_1 x_{i1} + b_2 \bar{x}_2)}, \quad i = 1, 2 \tag{7.8}$$

where $\bar{x}_2$ is the average value of log(WBC).

An alternative and arguably better method for constructing adjusted survival curves, suggested by Makuch (1982), is to take the average of the curves rather than the curve of the averages. If n is the total number of subjects, this method leads, for the present example, to the alternative formula

$$\bar{S}_{it} = \frac{1}{n} \sum_{k=1}^{n} S_{0t}^{\exp(b_1 x_{i1} + b_2 x_{i2})}, \quad i = 1, 2 \tag{7.9}$$

Figure 7.13 shows the model survival curves corresponding to each AG group, adjusted for log(WBC) using Equation (7.9).

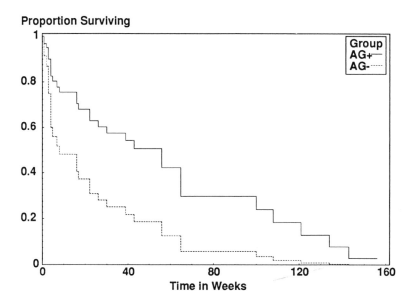

FIGURE 7.13: Model Survival Curves adjusted for WBC: Leukemia Patients

As a further illustration of modelling survival curves, consider the survival times of 137 patients with lung cancer cited in the book by Kalbfleisch and Prentice (1980). Here the outcome is survival time in days, and exposure variables include treatment ('test' or 'standard'), type of cell affected ('adeno' or other), and performance status using a scale from 10 (worst) to 99 (best).

Figure 7.14 shows Kaplan–Meier curves of the survival times for the two groups separated by cell type, which is known to be an important risk factor for lung cancer. Although the patients were followed up for almost three years, only 10 of them survived more than a year, so the graph is truncated after 400 days. The numbers at risk at regular intervals of time are given below the horizontal axis.

It is clearly evident from the survival curves that the 27 patients with adenocarcinomas had shorter survival times than the others: their median survival time is about half that of the other patients. Moreover the logrank test gives a p-value of 0.004.

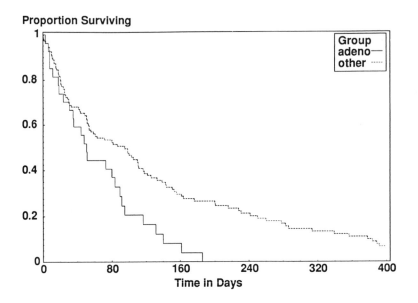

FIGURE 7.14: Kaplan–Meier Curves for Lung Cancer Patients

Now look at the printout shown in Table 7.14. This shows the result of fitting the model

$$h_{it} = h_{0t} \exp(b_1 x_{i1} + b_2 x_{i2} + b_3 x_{i3}) \qquad (7.10)$$

to the lung cancer survival data. The parameters b_1, b_2 and b_3 correspond to the predictors 'treatment', 'adeno cell' and 'performance status', and it is assumed that their effects are additive. Note that two of the p-values for testing the PH model are small, the one corresponding to performance status being extremely small, so the model does not provide an acceptable fit to the data.

```
Cox Regression Analysis
Response: Survival time (days)
Col Name        Coeff StErr p-val HazRat   95%    CI P(PH)
```

Col Name	Coeff	StErr	p-val	HazRat	95%	CI	P(PH)
3 Treatment	0.067	0.188	0.720	1.070	0.740	1.546	0.893
4 Adeno	0.720	0.240	0.003	2.054	1.283	3.289	0.051
5 Perf Stat	-0.033	0.005	0.000	0.967	0.958	0.977	0.001

```
n:137          %Cen:6.569        -2LogL:961.130        #iter:9
```

TABLE 7.14: Proportional Hazards Model: Lung Cancer Data

Three further methods for analysing survival data are described in the next section. These methods provide an alternative or more general approach to the analysis of survival data, and are worth trying when the proportional hazards assumption is not supported by the data. The first method is a simple extension of the linear regression model described in Chapter 3. Stratification is the basis for the other two methods.

7: Other Models

Censored Linear Regression

Since survival times t_i are measured on a continuous scale, the linear regression model

$$g(t_i) = b_0 + \sum_{j=1}^{p} b_j x_{ij} \qquad (7.11)$$

may be fitted, where $g(t)$ is some transformation of the survival times, such as the logarithm. The model may be fitted using maximum likelihood based on an assumption that the errors comprise a sample from a normal distribution.

Table 7.15 shows the result of fitting this model to the base 2 logarithms of the leukemia survival times given in Table 7.11. The base 2 logarithms are preferred because a unit increase in the transformed data scale corresponds to a doubling of the survival time. In this case none of the survival times is censored, so it is a straightforward matter to fit the linear regression model and examine residuals. Figure 7.15 shows the graph of residuals plotted against normal scores, indicating that the linear model fits these data well. The *p*-value from the Shapiro–Wilk test for normality is 0.76, so there is no evidence of a departure from normality.

```
Linear Regression Analysis
Response: base 2 logarithm of survival time (weeks)
  Column Name         Coeff   StErr p-value           SS
```

Column	Name	Coeff	StErr	p-value	SS
0	Constant	11.191	2.330	0.000	549.987
2	AG+	1.426	0.629	0.031	20.652
3	log(WBC)	-1.896	0.547	0.002	39.007

```
df:30         RSq:0.38       s:1.802       RSS:97.363
```

TABLE 7.15: Linear Regression: Leukemia Survival Times

It is interesting to compare the results obtained using the log-linear model with those given by the proportional hazards model (see

Table 7.12). The PH model states that the AG positive patients have a relative risk 0.36 compared with the others, whereas the log-linear model states that the AG positive patients have survival times which are on average $2^{1.43}$, or 2.69 times longer than the others. These results are quite consistent.

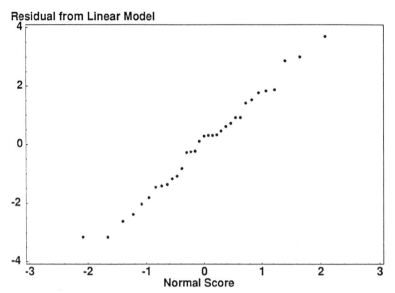

FIGURE 7.15: Normal Scores Plot – Residual (log) Leukemia Survival Times

The lung cancer survival times may also be modelled using linear regression, and the results are shown in Table 7.16. An acceptable fit to normality, indicated by the normal scores plot of residuals shown in Figure 7.16 (with censored data denoted by larger, lightly shaded circles), is obtained by choosing the transformation $\log(5 + t)$, where t is the survival time in days. Adding 5 days to the survival times before taking logarithms improves the fit substantially, because there are a few patients who had very short survival times who would otherwise have unduly influenced the results.

Table 7.16 shows that the test treatment is not a risk factor but the other variables are. The estimated coefficient −0.61 for adenocarcinoma cells means that there is a reduction by a factor $2^{-0.61}$, or 0.66 in survival duration associated with these patients. Similarly the model predicts an increase by a factor $2^{0.049} = 1.035$ in survival duration associated with each unit increase in the performance status score. These survival durations are all measured from a time origin 5 days before the patients entered the trial.

```
Linear Regression Analysis with Censored Data
Response: base 2 logarithm of days survived + 5
   Column Name                      Coeff   StErr p-value

        0 Constant                  3.487   0.388   0.000
        3 Test treatment           -0.036   0.238   0.880
        4 Adeno cell               -0.606   0.298   0.042
        5 Performance status        0.049   0.006   0.000

df:133        -2logL:448.478          s:1.354      #iter:6
```

TABLE 7.16: Linear Regression: Lung Cancer Patients

Residual: base 2 log survival time + 5 days

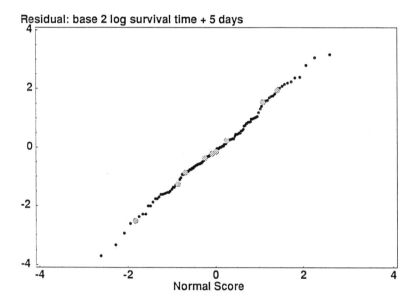

FIGURE 7.16: Normal Scores Plot of Residual Cancer Survival Times

Stratifying by Covariates

Judging by the small *p*-values for testing the PH hypothesis for cell type and performance status given in Table 7.14, the proportional hazards model does not fit the lung cancer survival data. A way of addressing the problem is fit separate PH models to the data stratified by cell type and performance status. However this approach suffers from the usual problems of subgroup analysis: if the subgroups are small, it will be more difficult to detect a treatment effect if it exists. What is needed is a kind of Mantel–Haenszel analysis that produces combined estimates of effects.

The stratified method involves fitting a statistical model in which there is a single set of parameter estimates corresponding to the covariates (as in the PH model), but instead of having a single base-line survival curve, each stratum has its own baseline survival curve. This model takes the form

$$h_{gt} = h_{g0t} \exp(\Sigma b_j x_j) \tag{7.12}$$

where g denotes the stratum.

Consider the application of the method to the lung cancer survival data. The printout from fitting separate PH models to the two cell types is given in Table 7.17. This printout again suggests that there is no overall treatment effect, but the p-value for testing the PH assumption with respect to performance status remains small, so the model is unacceptable.

```
Stratified Cox Regression Analysis on Cell Type
Response: Survival Time in Days
Col Name        Coeff StErr p-val HazRat    95%    CI P(PH)
```

Col Name	Coeff	StErr	p-val	HazRat	95%	CI	P(PH)
3 Treatment	0.098	0.189	0.604	1.103	0.762	1.596	0.652
5 Perf Stat	-0.036	0.005	0.000	0.964	0.954	0.974	0.002

```
n:137       %Cen: 6.569      -2LogL: 837.393         #iter:7
```

TABLE 7.17: PH Model Stratified by Cell Type: Lung Cancer Data

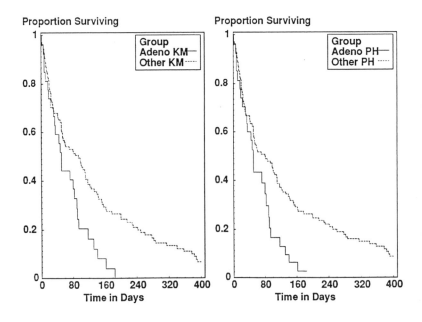

FIGURE 7.17: Kaplan–Meier and Model Curves: Lung Cancer Patients

The right-hand panel of Figure 7.17 shows adjusted survival curves (using Equation (7.9)) for the two cell types based on the stratified model. For purposes of comparison the unadjusted (Kaplan–Meier) curves from Figure 7.14 are repeated in the left-hand panel of the figure. You can see that the curves are virtually indistinguishable from each other. However, in view of the poor fit of the model these results are not particularly useful.

A more effective method of stratification, based on dividing the time axis into zones, is described next.

Stratifying by Time

You have seen an example where a violation of the proportional hazards assumption for a covariate is tackled by fitting a more general model in which separate baseline hazard functions are assumed for different strata. This method is useful whenever the PH assumption is violated for a covariate (or a set of covariates) different from the primary risk factor of interest. However, if the risk factor of interest fails the PH assumption, stratifying with respect to it means that there are no longer model parameters encapsulating the effect of this risk factor, but rather model curves. For example, if cell type were the primary risk factor for survival in the lung cancer trial, its effect would be summarised in the right-hand panel of Figure 7.17 rather than in the printout in Table 7.17.

If there are no additional covariates, as in the study involving the hepatitis patients (see Table 7.10), stratifying on the risk factor of interest simply gives the Kaplan–Meier curves. Thus another modelling approach is needed when the PH assumption fails for the primary risk factor.

Stratifying with respect to time is an alternative method. In this method the time axis is divided into non-overlapping zones and separate PH models are fitted within each time zone.

The hepatitis study provides a simple illustration. Taking two time zones separated at 70 months gives the result in Table 7.18. The treatment effect is statistically significant only in the first time zone, indicating that the treatment effect wears off after about six years.

Survival curves may be computed from a time-stratified model by piecing together the model-based curves in the different time zones as described by Harrell and Lee (1986). Figure 7.18 shows the graphs corresponding to the model reported in Table 7.18. These graphs should be compared with those shown in Figure 7.7. You can see that the time stratified PH model gives an improved fit.

```
Time-stratified Cox Regression Analysis
Response: Survival time (months)
 Name   Zone   Coeff   StErr   p-val   HazRat    95%     CI
```

Name	Zone	Coeff	StErr	p-val	HazRat	95%	CI
Pred:	0-70	-1.429	0.520	0.006	0.239	0.086	0.664
Pred:	70+	0.974	1.083	0.369	2.648	0.317	22.136

```
 n:44       %Cen: 38.636      -2LogL: 168.008       #iter:7
```

TABLE 7.18: PH Model Stratified by Time Zone: Hepatitis Patients

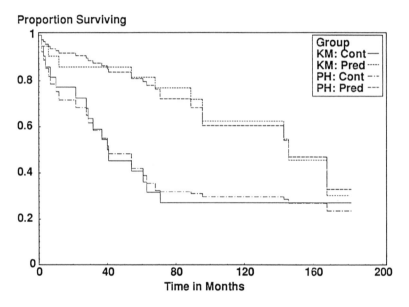

FIGURE 7.18: Time-stratified PH Model Survival Curves: Hepatitis Patients

As a second example let us return to the lung cancer survival times. Table 7.19 shows the printout obtained from fitting a time-stratified PH model with the three zones 0–49 days, 50–99 days, and 100 or more days. There is some evidence that the treatment may have a protective effect after 100 days but not before this duration, that the adenocarcinoma cell type becomes a risk factor after 50 days, and that the effect of performance status wears off after 100 days.

```
Time-stratified Cox Regression Analysis
Response: Survival time (days)
Name        Zone   Coeff StErr p-val HazRat   95%     CI
```

Name	Zone	Coeff	StErr	p-val	HazRat	95%	CI
Treat:	0-50	0.081	0.298	0.786	1.084	0.604	1.946
Treat:	50-100	0.666	0.418	0.110	1.947	0.859	4.414
Treat:	100+	-0.788	0.344	0.022	0.455	0.232	0.893
Adeno:	0-50	0.135	0.340	0.691	1.145	0.588	2.229
Adeno:	50-100	1.084	0.449	0.016	2.955	1.226	7.124
Adeno:	100+	1.033	0.524	0.048	2.810	1.007	7.842
PerfS:	0-50	-0.049	0.007	0.000	0.952	0.939	0.965
PerfS:	50-100	-0.040	0.013	0.002	0.961	0.937	0.985
PerfS:	100+	0.005	0.011	0.665	1.005	0.983	1.028

```
n:137      %Cen: 6.569      -2LogL: 934.340      #iter:7
```

TABLE 7.19: PH Model Stratified by Time Zone: Lung Cancer Patients

Summary

In this chapter methods for survival analysis have been introduced, as a further development of the methods for analysing follow-up studies considered in earlier chapters. In these methods the response variable is taken to be the interval of time that elapses before the event of interest occurs (called the 'failure' time), and there is a need to distinguish between complete intervals (when the event of interest occurs) and censored intervals (when the event has not yet been observed.)

For further discussion of models used to analysis survival data, the recent book by Collett (1994) is recommended.

The Survival Curve

The survival curve is the proportion of subjects surviving beyond a given duration of time, and is used to compare the prospects of different groups of subjects in a follow-up study. The Kaplan–Meier curve is the estimated survival curve based on a sample of subjects' survival times.

Comparing Survival Curves

Two survival curves may be compared by stratifying the data by failure time and using the Mantel–Haenzsel method to adjust the incidence density ratio. Two or more survival curves may be compared using a test such as the logrank test, or by using Poisson regression with failure time as a stratification variable. Using treated and control groups for comparison, the methods are illustrated with remission times of leukemia patients and survival times of patients with chronic hepatitis.

The Proportional Hazards Model

The proportional hazards model specifies the risk of the event occurring to an individual subject as the product of a function of time, common to all subjects, and an exponential linear function of nominated risk factors. This model has a similar interpretation to Poisson regression and is easily fitted to case-by-case survival data. The validity of the proportional hazards model may be assessed graphically or tested giving a p-value for each covariate.

Modelling Covariates

In common with other multiple regression models, proportional hazards regression can handle multiple risk factors or covariates, and can also provide estimates of survival curves adjusted for specified covariates. We considered two further examples: an observational study of leukemia patients classified by characteristics and counts of white blood cells, and a clinical trial of lung cancer patients with treatment, cell type and performance status as risk factors.

Other Models

We discussed two methods for coping with departures from the proportional hazards model. The first method involves fitting an ordinary linear regression model to the survival times, appropriately transformed, while the second involves stratification, either (a) with respect to offending covariates, or (b) by dividing the time interval into zones and fitting and reassembling separate PH model within these zones.

Exercises

Exercise 7.1: Analyse the data from the motion sickness experiment given in Exercise 4.3 using the methods in Sections 3 and 4 of this chapter.

Exercise 7.2: Analyse the departure times of the psychiatric patients (considered in Section 6 of Chapter 6) by fitting an appropriate proportional hazards model, and compare your results with those obtained from the logistic regression model fitted in Chapter 6. Which method is preferable? Give reasons for your answer.

Exercise 7.3: The following data are survival times in weeks of patients with lymphocytic non-Hodgkins lymphoma, classified into two groups (see Dinse (1982) and Kimber (1990)). The asterisks denote censored survival times. Compare the two populations using the methods outlined in Sections 3 of 4 of this chapter.

Asymptomatic			Symptomatic
50	257	349*	49
58	262	354*	58
96	292	359	75
139	294	360*	110
152	300*	365*	112
159	301	378*	132
189	306*	381*	151
225	329*	388*	276
239	342*	281	
242	346*	362*	

Exercise 7.4: Anderson et al (1980, page 218) augmented the leukemia remission times (Table 7.2) by adding a hypothetical white blood count, the logarithms of which (in the same order as Table 7.2) are as follows. Analyse these data by the methods given in Section 6 of this chapter.

Treatment Group: 2.31 4.06 3.28 4.43 2.96 2.88 3.60 2.32 2.57 3.20 2.80
                           2.70 2.60 2.16 2.05 2.01 1.78 2.20 2.53 1.47 1.45

Control Group: 2.80 5.00 4.91 4.48 4.01 4.36 2.42 3.49 3.97 3.52 3.05
                          2.32 3.26 3.49 2.12 1.50 3.06 2.30 2.95 2.73 1.97

Exercise 7.5: The following data were reported by Holt and Prentice (1974) and examined further by Woolson and Lachenbruch (1980). They are survival times in days of closely matched and poorly matched skin grafts on the same burn patients, with the asterisks denoting censoring.

(a) Compare the survival times in the two populations using both a paired *t*-test and a nonparametric test. In each case treat the two censored observations as if they were complete data. Do you think it is reasonable to do this? Why?

(b) Compare the survival times in the two populations using the methods in Sections 3 and 4 of the present chapter. In this case ignore the fact that two observations are taken from the same subject, that is, treat the data as if they comprised two independent samples of 11 survival times. Do you think it is reasonable to do this? Why?

(c) Based on the two analyses, what is your conclusion?

Patient	Close match	Poor match
1	37	29
2	19	13
3	57*	15
4	93	26
5	16	11
6	22	17
7	20	26
8	18	21
9	63	43
10	29	15
11	60*	40

Exercise 7.6: In 1982 the Gastrointestinal Tumor Study Group in the US reported results from a study comparing chemotherapy with combined chemotherapy and radiotherapy in the treatment of locally unresectable gastric cancer, and the survival times in days, listed below, were compared by Stablein and Koutrouvelis (1985). (The censored data are denoted with + symbols.) Fit a time-stratified proportional hazards model to these data.

Chemotherapy Alone

1	63	105	129	182	216	250	262	301
301	342	354	356	358	380	383	383	388
394	408	460	489	499	523	524	535	562
569	675	676	748	778	786	797	955	968
1000	1245	1271	1420	1551	1694	2363	2754+	2950+

Combined Therapy

17	42	44	48	60	72	74	95	103
108	122	144	167	170	183	185	193	195
197	208	234	235	254	307	315	401	445
464	484	528	542	567	577	580	795	855
1366	1577	2060	2412+	2486+	2796+	2802+	2934+	2988+

Exercise 7.7: Analyse the leukemia remission times (Table 7.2) using (a) logistic regression (where the outcome event is defined as a remission within nine weeks), and (b) Poisson regression (where the outcome event is remission and the person-weeks of follow-up is taken into account. Compare these results with that obtained using the proportional hazards model.

Exercise 7.8: The following table shows the graft survival times (*t*) in months, together with the total number of HLA-B or DR antigen mismatches between donor and recipient (*x*), for 148 renal transplant patients reported by Henderson and Milner (1991, Table 1). (The censored survival times are denoted by asterisks.) Analyse these data using survival analysis techniques.

t	x	t	x	t	x	t	x	t	x
0.035*	3	3.803	3	12.213*	1	19.508	2	32.672	1
0.068*	0	4.311	1	12.508*	3	19.574	3	32.705	2
0.100*	0	4.867	0	12.533	2	19.733	0	33.148	1
0.101*	1	5.180*	1	13.467	0	20.148	2	33.567	1
0.167	4	6.233	2	13.800	2	20.180	0	33.770	1
0.168*	2	6.367	2	14.267	0	20.900*	2	33.869	2
0.197	1	6.600	1	14.475	4	21.167	0	34.836	0
0.213*	1	6.600	0	14.500	1	21.233	0	34.869	1
0.233	1	7.180*	3	15.213	1	21.600	3	34.934	2
0.234*	2	7.667	1	15.333	0	22.100	1	35.738	0
0.508*	0	7.733*	1	15.525	1	22.148	2	36.180	1
0.508	2	7.800	2	15.533	2	22.180	0	36.213	1
0.533*	3	7.933	1	15.541	1	22.180	0	39.410	1
0.633	0	7.967	1	15.934	0	22.267	0	39.433	0
0.767*	3	8.016*	2	16.200	1	22.300	2	39.672	0
0.768*	4	8.300*	1	16.300	0	22.500	1	40.001	0
0.770	0	8.410	0	16.344	1	22.533	1	41.733	2
1.066*	4	8.607	1	16.600	0	22.867	1	41.734	0
1.267	2	8.667*	1	16.700	1	23.738	1	42.311	2
1.300*	3	8.800	1	16.933	3	24.082	1	42.869	0
1.600*	1	9.100	0	17.033	3	24.180	0	43.180	0
1.639	2	9.233*	1	17.067	0	24.705	0	43.279	1
1.803	2	10.541	2	17.475	1	25.705	2	43.902	2
1.867*	4	10.607	3	17.667	1	25.213	1	44.267	2
2.180*	3	10.633	1	17.700	1	29.705	3	44.475	1
2.667*	4	10.667*	2	17.967	1	30.443	1	44.900	1
2.967	1	10.869	3	18.115	2	31.667	0	45.148	1
3.328	2	11.067*	2	18.115	1	31.934	2	46.451	0
3.393*	3	11.180	0	18.933	0	32.180	1		
3.700*	4	11.443	0	18.934	1	32.367	0		

Exercise 7.9: For the data listed in Exercise 2.3 (steady-state haemoglobin levels for patients with sickle cell disease), compare the means in the three groups using linear regression analysis. Are the statistical assumptions reasonable?

Now repeat the analysis using the proportional hazards model instead of the linear regression model, and compare the conclusion with that obtained from the regression analysis.

References

Anderson, S., A. Auquier, W.W. Hauck, D. Oakes, W. Vandaele & H.I. Weisberg (1980): *Statistical Methods for Comparative Studies*. John Wiley & Sons. New York.

Collett, D. (1994): *Modelling Survival Data in Medical Research*. Chapman & Hall. London.

Cox, D.R. (1972): Regression models and life-tables (with discussion), *Journal of the Royal Statistical Society (B)*, **34**, pages 187–220.

Dinse, G.E. (1982): Nonparametric estimation for partially complete time and type of failure data, *Biometrics*, **36**, pages 417–431.

Feigl, P. & M. Zelen (1965): Estimation of exponential survival probabilities with concomitant information, *Biometrics*, **21**, pages 826–838.

Freireich, E.J. et al (1963): The effect of 6-mercaptopurine on the duration of steroid-induced remissions in acute leukemia, *Blood*, **21**, pages 699–716.

Harrell, F. & K. Lee (1986): Verifying assumptions of the proportional hazards model, *Proceedings of the Eleventh Annual SAS User's Group International*, pages 823–828.

Henderson, R. & A.Milner (1991): Aalen plots under proportional hazards, *Applied Statistics*, **40**, pages 401–409.

Holt, J.D. & R.L. Prentice (1974): Survival analysis in twin studies and matched pairs experiments, *Biometrika*, **61**, pages 17–30.

Kalbfleisch, J. & R.L. Prentice (1980): *The Statistical Analysis of Failure Time Data*. John Wiley & Sons. New York.

Kaplan, E.L. & P. Meier (1958): Nonparametric estimation from incomplete observations, *Journal of the American Statistical Association*, **53**, pages 457–481.

Kimber, A.C. (1990): Exploratory data analysis for possibly censored data from skewed distributions, *Applied Statistics*, **39**, pages 21–30.

Kirk, A.P. et al (1980): Late results of the Royal Free Hospital prospective controlled trial of prednisolone therapy in hepatitis B surface antigen negative chronic active hepatitis, *Gut*, **21**, pages 78–83.

Makuch, R.W. (1982): Adjusted survival curve estimation using covariates, *Journal of Chronic Diseases*, **35**, pages 437–443.

Pocock, S.J. (1986): *Clinical Trials: A Practical Approach.* John Wiley & Sons. Chichester.

Prentice, R.L. & P. Marek (1979): A qualitative discrepancy between censored rank tests, *Biometrics*, **35**, pages 861–867.

Schoenfeld, D. (1982): Partial residuals for the proportional hazards model, *Biometrika*, **69(1)**, pages 239–241.

Stablein, D.M & I.A. Koutrouvelis (1985): A two-sample test sensitive to crossing hazards in uncensored and singly censored data, *Biometrics*, **41**, pages 643–652.

Woolson, R.F. & P.A. Lachenbruch (1980): Rank tests for censored matched pairs, *Biometrika*, **67**, pages 597–606.

8

MATCHING

1: Introduction

In the preceding chapters you have seen the basic statistical methods for analysing data from epidemiological studies. These methods involve analysing data with a view to quantifying associations between outcome events and possible risk factors, including exposures and stratification variables. Various statistical models may be used, including logistic, Poisson, and proportional hazards regression. These models take account of confounding variables and they apply to all study designs.

The present chapter introduces the concept of *matching*. In Chapter 1 bias and sampling variability were given as two factors that can reduce the credibility of an epidemiological study. Matching is a study design technique aimed at reducing these effects.

The simplest instance of matching is pair-matching, where the units considered in the study are not individual items but pairs of components with something in common. The two components are selected to have either different exposures or different outcomes (depending on the study type). If the components are individual subjects they could be twins or siblings: studies involving genetically similar individuals play an important role in epidemiological research.

As an example, consider an hypothetical study to investigate the effect of an immunisation treatment program on the incidence of sickness among schoolchildren. A possible protocol would involve choosing a sample from the target population of interest, dividing it into two similar groups (using a randomised allocation to avoid selection bias), giving the treatment to one group and a placebo to the other, and recording the number of days of sickness during the following year for each child. Suppose the data given in Table 8.1 were obtained from 20 subjects. The sample means are 4.6 days in the control group and 3.1 in the treated group, so the estimated treatment benefit is 1.5 days, with standard error (Equations (2.5) and (2.6)) 0.985. An approximate 95% confidence interval, based on two standard errors on each side of the estimate, is thus (−0.47, 3.47) days. This interval includes 0, so the treatment could be ineffective. Moreover the two-sample t-test of the null hypothesis that the mean number of days of sickness is the same in the treated and untreated populations gives a nonsignificant p-value (0.145).

Control Group	Treated Group
4	6
8	3
5	2
4	0
7	3
0	2
6	6
2	3
7	2
3	4

TABLE 8.1: Days of Sickness in 20 (hypothetical) Schoolchildren

Now consider an alternative study design in which the experimental units are 10 twins instead of 20 independent individuals. One twin (randomly chosen) from each pair is treated and the other is not. The total number of days of sickness during the following year is again recorded for each child, giving the data shown in Table 8.2. Note that these data are the same as those given in Table 8.1, except that the outcomes in the treated group are ordered differently.

Pair	Control	Treated	Control – Treated
1	4	0	4
2	8	6	2
3	5	3	2
4	4	2	2
5	7	6	1
6	0	2	-2
7	6	3	3
8	2	2	0
9	7	4	3
10	3	3	0

TABLE 8.2: Days of Sickness in ten (hypothetical) Twins

Now the data in Table 8.2 may be regarded simply as one sample of ten observations, each observation being the number of additional days of sickness experienced by the untreated twin in each pair: these are given as the right-hand column of Table 8.2. These observations have a sample mean of 1.5 days, and assuming normality a 95% confidence interval for the treatment benefit, based on Equation (2.4), is (0.23, 2.77) days, indicating that the treatment is beneficial. The p-value for testing the null hypothesis of no treatment effect, based on a t-test with 9 degrees of freedom, is 0.026.

This example shows that a study based on matched pairs can be more powerful than an equivalent study based on independent subjects. An advantage of the paired design is that the statistical sampling variation is reduced, possibly quite substantially. It is reasonable to assume that there is less variation between the levels of sickness experienced by children in the same family than by children from different families.

The matched pairs design may be regarded as a set of small studies, each involving just two subjects. If the two subjects are similar and they are given different treatments, the outcome difference may be ascribed to the treatment effect. These within-pair treatment effects may then be aggregated to estimate the overall treatment benefit in a population.

The data pairs could be measured on the same subject. For example you could compare the efficacies of different skin graft treatments by applying each treatment to the same person. Even in cases where the individual must be treated as a whole, it is possible for the two members of the matched pair to comprise the same individual. In crossover studies each subject is given the treatments sequentially. Here it is assumed that the outcome is reversible, so that, after a suitable recovery period, the individual receives the second treatment in the same state as when the first treatment was applied.

Designs involving matched pairs may be generalised in various ways. If more than two treatments or risk factors need to be compared, you could form matched sets of more than two similar components. If the sets contain several similar components but you only wish to compare a smaller number of treatments, a stratified design may be used.

If the outcomes are continuous, standard statistical methods including multiple regression and the analysis of variance may be used to analyse matched data. In epidemiological research it is more usual to have dichotomous outcomes, and the present chapter covers methods for analysing matched dichotomous data.

First we consider the problem of estimating an odds ratio from matched data where the risk factor is dichotomous, and methods for matched analysis for both case-control and cohort studies are given. The logistic model is then introduced as a method for handling covariates, for designs involving both matched pairs and matched sets containing more than one control or nonexposed subject, and it is shown that conditional logistic regression models are appropriate. Before-after studies are considered as a special case of pair matched studies. The chapter concludes with some discussion of the limitations of matching in epidemiological studies.

2: Matched Pairs

Matched pairs may arise from a cross-sectional, cohort, or case-control design. As an example of a case-control design, consider the following study undertaken by Ms Skulrat Rithsmithchai at Prince of Songkla University in 1993. Here the outcome is pre-term birth (less than 37 weeks gestation) and the risk factor of interest is whether the mother's predominant work activity during the first three months of pregnancy involved standing (rather than sitting, walking, or some combination of activities). The study comprised 223 pairs of women who had given birth at the local regional hospital during 1992 and 1993. One member of each pair (the *case*) had delivered a pre-term baby and the other (the *control*) had not, and these pairs were matched for maternal age and parity.

There are four possible exposure combinations for each pair: (1) both case and control exposed, (2) case exposed and control non-exposed, (3) case nonexposed and control exposed, and (4) both case and control nonexposed. Denoting the numbers of outcomes in these four categories by n_{11}, n_{10}, n_{01} and n_{00}, respectively, the results from the study may thus be presented in a 2-by-2 table as follows.

		Control	
		E	$\overline{E}$
Case	E	n_{11}	n_{10}
	$\overline{E}$	n_{01}	n_{00}

TABLE 8.3: Presentation of Data from a Matched Case-control Study

For Ms Skulrat's study the numbers shown in Table 8.4 were obtained. There seems to be some evidence that the exposure is associated with the outcome, since there were 31 pairs in which the exposed member experienced the outcome and the nonexposed member did not, compared with only 14 pairs in which these outcomes were reversed.

| | | Full-term Birth (Control) | |
		Standing	Not Standing
Pre-term Birth (Case)	Standing	177	31
	Not Standing	14	1

TABLE 8.4: Data from a Study of 223 Matched Pairs of Thai Mothers

Consider estimating the odds ratio from matched data as given in Table 8.3. If we assume that an odds ratio cannot be estimated from a study in which all the subjects have the same exposure level, then the matched pairs in which the case and the control are either both exposed, or both nonexposed, contribute no information about the odds ratio. It follows that the relevant information is confined to the *discordant* pairs.

The discordant pairs have one member of each pair exposed to the risk factor and the other nonexposed. Suppose that p_1 and p_0 represent the probabilities of the outcome occurring to an exposed and a nonexposed individual, respectively. Then the probability that the exposed member of a discordant pair experiences the outcome and the nonexposed member does not is $p_1(1-p_0)$. Similarly the probability that the exposed member does not experience the outcome and the non-exposed member does is $(1-p_1)p_0$. The ratio of these probabilities is just the odds ratio O_1/O_0, where $O = p/(1-p)$ is the odds associated with an outcome risk p. Thus the odds ratio in a matched case-control study may be estimated simply by taking the ratio of the number of positive to negative discordant pairs, that is,

$$OR = \frac{n_{10}}{n_{01}} \tag{8.1}$$

An asymptotic formula for the standard error of the natural logarithm of this estimate is

$$SE[\ln(OR)] = \sqrt{\frac{1}{n_{10}} + \frac{1}{n_{01}}} \tag{8.2}$$

It turns out that Equations (8.1) and (8.2) are special cases of Equations (4.7) and (4.8), so they are equivalent to the Mantel–Haenszel formulas for estimating the odds ratio from a stratified case-control study. For the matched pairs design, each stratum contains just one case and one control subject.

An approximate 95% confidence interval for the odds ratio may be obtained from Equations (8.1) and (8.2) in the usual way, that is, by taking two standard deviations on each side of the estimated log odds ratio and exponentiating the result. Thus for the data from Table 8.4 you obtain an estimated odds ratio $31/14 = 2.21$, an estimate 0.322 for the standard error of the log odds ratio, and a 95% confidence interval (1.16, 4.22), indicating that standing while working is a risk factor for preterm birth in the target population of Thai women.

The null hypothesis of no association between the exposure and outcome may be tested using a chi-squared statistic (with 1 degree of freedom), as suggested by McNemar (1947). The formula is

$$\chi^2_{McN} = \frac{(n_{10} - n_{01})^2}{n_{10} + n_{01}} \tag{8.3}$$

It may be shown that this test statistic is identical to the Mantel–Haenszel test statistic given by Equation (4.6). For the data from Table 8.4, the value is $17^2/45 = 6.42$ corresponding to a p-value 0.011.

Matched versus Non-matched Analysis

The four possible outcomes in a matched pairs case control study giving rise to Table 8.3 may also be represented as follows:

Category	Count
DE, $\bar{D}E$	n_{11}
DE, $\bar{D}\bar{E}$	n_{10}
$D\bar{E}$, $\bar{D}E$	n_{01}
$D\bar{E}$, $\bar{D}\bar{E}$	n_{00}

Ignoring the matching, you may construct a 2-by-2 table showing the association between the exposure and the outcome, as follows:

		Exposure	
		E	$\bar{E}$
Outcome	D	$n_{11}+n_{10}$	$n_{01}+n_{00}$
	$\bar{D}$	$n_{11}+n_{01}$	$n_{10}+n_{00}$

TABLE 8.5: Unmatched Data from a Matched Case-control Study

This table may be used as a basis for estimating the odds ratio and its confidence interval using the methods outlined in Chapters 4 and 5. However, these methods ignore the matching and thus may give biased estimates.

Consider the data from Table 8.3. Ignoring the matching gives the data in Table 8.6.

		Exposure	
		Standing	*Not Standing*
Birth Outcome	*Pre-term*	208	15
	Full-term	191	32

TABLE 8.6: Unmatched Data from Study of 446 Thai Mothers

Using Equations (4.1) and (4.2) estimates $(208 \times 32)/(15 \times 191) = 2.32$ and 0.329 are obtained for the odds ratio and the standard error of its logarithm, giving a 95% confidence interval (1.22, 4.42). These estimates are very similar to those obtained from Equations (8.1) and (8.2), which take the matching into account.

The size of the discrepancy between the matched and the unmatched estimates of the odds ratio in a matched case-control study depends on the degree of association within the matched pairs. To see this, suppose that Ms Skulrat had obtained the following results for her study instead of those given in Table 8.4.

		Full-term Birth (Control)	
		Standing	*Not Standing*
Pre-term Birth (Case)	*Standing*	147	31
	Not Standing	14	31

TABLE 8.7: Hypothetical Data from 223 Matched Pairs of Thai Mothers

Note that these data give precisely the same estimate and confidence interval for the odds ratio as the original data, since the numbers of discordant pairs in the two categories are the same as before. However, if the matching is ignored Equations (4.1) and (4.2) give an estimate $(178 \times 62)/(45 \times 161) = 1.52$ for the odds ratio and 0.224 for the standard error of its logarithm, and a 95% confidence interval (0.98, 2.36). These values are quite different from those given by Equations (8.1) and (8.2) based on the matched analysis.

The reason for the difference is that the unmatched analysis ignores any correlation in exposure status between the case and the control within the matched pairs. If this correlation is substantial, the unmatched analysis may give a biased result. Moreover the extent of this correlation may be assessed by examining Table 8.3: if there is no association between the exposure status of a case and its matched control, the counts in Table 8.3 should factorise and the odds ratio computed from this table should be close to 1 (or at least statistically indistinguishable from 1).

As a check, for the data in Table 8.4 Equations (4.1) and (4.3) give an estimated odds ratio $(177 \times 1)/(31 \times 14) = 0.41$ and a 95% confidence interval (0.05, 3.21), showing no evidence of correlation. In contrast, the equivalent odds ratio estimated from the data in Table 8.7 is 10.5 with a 95% confidence interval (5.01, 22.02), indicating a strong within-pair correlation. This explains why the matched and unmatched analyses give such different results for the data in Table 8.7.

If there is no evidence of correlation within pairs, should the matched analysis still be undertaken? Not necessarily. Although the matched analysis is more appropriate if the data are matched it can give unstable results when the sample size is too small. For example if either n_{10} or n_{01} is 0, Equation (8.2) does not give a finite standard error estimate, but Equation (4.2) still gives a finite estimate provided n_{11} and n_{00} are positive.

As a further illustration of a pair-matched case-control study, consider the following data from a study undertaken by Donovan et al (1984). In this study 8502 pairs of new-born Australian babies were matched by hospital, date of birth, and age and medical insurance status of mother. The cases were born with certain birth defect anomalies and the controls were born without these anomalies, and the exposure was the father having been a soldier with the Australian army during the Vietnam war (and thus at risk of exposure from the 'agent orange' defoliant which is suspected to cause birth defects).

		No Birth Anomaly (Control)	
		Vietnam	non-Vietnam
Birth Anomaly (Case)	Vietnam	2	125
	non-Vietnam	121	8254

TABLE 8.8: Data from 'Agent Orange' Study

Using Equation (8.1) an estimate $125/121 = 1.033$ is obtained for the odds ratio and Equation (8.2) gives 0.128 for the estimated standard error of its logarithm, giving a 95% confidence interval (0.80, 1.33). These estimates are almost exactly the same as those obtained from the unmatched analysis: using Equation (4.1) the estimated odds ratio is $(127 \times 8375)/(123 \times 8379) = 1.032$, and Equation (4.2) gives a standard error 0.127, so a 95% confidence interval is (0.80, 1.32). As for the preceding example, there is no evidence in Table 8.8 of within pair correlation.

You may conclude from this study, as the authors did, that there is no evidence of an association between exposure to the defoliant and an increased risk of fathering a child with anomalies. However, the upper limit of the confidence interval indicates that the odds ratio could be as high as 1.32, so the study is not really conclusive.

Matched Cohort Studies

Although Equations (8.1) and (8.2) have been derived for pair-matched case-control studies, these formulas, as well as McNemar's test (8.3) also apply to pair-matched cross-sectional or cohort studies with binary outcomes. In this case each matched pair contains one exposed and one nonexposed individual, and the data may be classified as follows:

Category	Count
DE, $\overline{D}$E	n_{11}
DE, $\overline{D}\overline{E}$	n_{10}
$\overline{D}$E, $\overline{D}$E	n_{01}
$\overline{D}$E, $\overline{D}\overline{E}$	n_{00}

These data from these matched pairs may be presented in the following contingency table, which is analogous to Table 8.3.

		Nonexposed Member	
		D	$\overline{D}$
Exposed Member	D	n_{11}	n_{10}
	$\overline{D}$	n_{01}	n_{00}

TABLE 8.9: Presentation of Data from a Matched Cohort Study

Cohort studies may be matched with respect to risk factors which are difficult to quantify, thus reducing unwanted variability in outcomes. In an ideal experiment comparing two treatments, the treatments are applied to pairs of subjects that are as alike as possible, so that differences in outcome are due to treatment effects. As an illustration, consider a study reported by Sleigh et al (1982) to compare the sensitivities of two methods (Kato–Katz and Bell) for detecting eggs from faeces samples. Each of 315 samples was divided into two halves, and the halves were tested using the different methods, giving the results shown in Table 8.10. The + and − symbols denote the outcomes eggs detected and no eggs detected, respectively.

		Kato-Katz	
		+	−
Bell	+	184	54
	−	14	63

TABLE 8.10: Comparison of Egg Detection Methods

Using Equation (8.1) the estimated odds ratio of detection for the Bell method compared with the Kato–Katz method is $54/14 = 3.86$, and Equation (8.2) gives a 95% confidence interval (2.68, 5.55), so the Bell method is clearly the more sensitive of the two.

Note that in this study ignoring the matching gives a quite different odds ratio estimate of 1.83 with 95% confidence interval (1.30, 2.58). This is not surprising since the outcomes within the matched pairs are highly correlated: Table 8.10 gives a estimate 15.3 (95% confidence interval 7.98–29.5) for the relative odds of getting a positive result in one half given a positive result in the other.

A further example is taken from Walker (1982), who reported some results from a study of 4830 pairs of men matched by year of birth and calendar time of follow-up, with one man from each pair given a vasectomy operation. Here the outcome of interest is the occurrence of a non-fatal myocardial infarct within the follow-up period. The results are summarised below in Table 8.11.

		No Vasectomy	
		MI	No MI
Vasectomy	MI	0	20
	No MI	16	4794

TABLE 8.11: Matched Cohort Studt of Vasectomy and Myocardial Infarct

Equations (8.1) and (8.2) give an estimated odds ratio 1.25 and 95% confidence interval (0.65, 2.41), so there is no evidence of an association between having a vasectomy and suffering a myocardial infarct. Moreover the confidence interval is very wide so the study is not particularly powerful. Note that there is no evidence of any association in outcome within the pairs, and the unmatched analysis gives almost exactly the same estimate and confidence interval.

In the next section the logistic model is introduced as a method for analysing matched binary data. As you saw in Chapter 5, a major advantage of this model is that covariates can be accommodated.

3: Logistic Modelling

As an illustration of likelihood theory, consider the simple situation involving unmatched binary data, where a subject i has outcome d_i (coded as 1 if positive and 0 otherwise) and exposure x_i. According to the logistic model (5.2), the likelihood of the outcome for subject i is

$$\frac{1}{1 + e^{-(a + bx_i)}}$$

if d_i is 1, and

$$\frac{e^{-(a + bx_i)}}{1 + e^{-(a + bx_i)}}$$

if d_i is 0. Combining the contributions from all n subjects, the likelihood of the observed data is thus given by the product

$$L = \prod_{i=1}^{n} \left(\frac{1}{1 + e^{-(a + bx_i)}} \right)^{d_i} \times \left(\frac{e^{-(a + bx_i)}}{1 + e^{-(a + bx_i)}} \right)^{1 - d_i} \tag{8.4}$$

Now consider the likelihood for data arising from a matched case-control study. For a given case-control pair indexed by i, let x_i denote the exposure for the case and let y_i denote the exposure for the control. Before the outcome has been observed, the chance that a subject with exposure x_i gets the outcome and a (matched) subject with exposure y_i does not, based on the logistic model (Equation (5.2)), is

$$Prob[D\bar{D} \mid x_i, y_i] = \frac{1}{1 + e^{-(a + bx_i)}} \times \left(1 - \frac{1}{1 + e^{-(a + by_i)}} \right)$$

Given that the exposure precedes the outcome, the situation *could* have been reversed, that is, a subject with exposure y_i could conceivably have experienced the outcome (and thus have become a case rather than a control); if so, the matched subject with exposure x_i would not have experienced the outcome (and consequently would have become a control instead of a case). Again using the logistic model, this probability is

$$Prob[\bar{D}D \mid x_i, y_i] = \left(1 - \frac{1}{1 + e^{-(a+bx_i)}}\right) \times \frac{1}{1 + e^{-(a+by_i)}}$$

Since the study involves only pairs in which one member is a case and the other is a control, the likelihood that the pair with joint exposure (x_i, y_i) becomes a case-control pair (rather than a control-case pair) is thus the ratio of the first expression to their sum, that is, after some cancellation,

$$L_i = \frac{1}{1 + e^{-b(x_i - y_i)}}$$

Assuming there are n case-control pairs in the study, each making an independent contribution, the likelihood associated with the observed data is the product of these components, that is

$$L = \prod_{i=1}^{n} \left(\frac{1}{1 + e^{-b(x_i - y_i)}}\right) \tag{8.5}$$

Given data (x_i, y_i), $i = 1, 2, ..., n$, the likelihood L may be maximised with respect to the parameter b, and the value of b which maximises this likelihood is known as the *maximum likelihood* estimator.

Comparing equations (8.4) and (8.5), you can see that (8.5) is the special case of (8.4) which arises when d_i is 1 for all values of i, the exposure for subject i is $x_i - y_i$ instead of just x_i, and a is equal to 0. It follows that you can estimate the parameter b in a matched case-control study simply by fitting a logistic model, without the constant term, to an appropriate data layout. In this data layout the outcomes are all positive and the exposures are defined as the *differences* between the case and control exposures within each matched pair.

As an illustration, consider again the data from the 'agent orange' study (Table 8.8). Following the prescription described above, the data layout has four records corresponding to the four cells in the table as follows. The printout shown below in Table 8.12 is obtained after fitting the logistic regression model to these data, with the constant term omitted from the model.

D	Total	E
2	2	0
121	121	−1
125	125	1
8254	8254	0

```
Logistic Regression Analysis
Response: D
  Col Name     Coeff   StErr p-value  Odds    95%    CI

    3 E        0.032   0.128      0.8 1.033  0.804 1.326
```

TABLE 8.12: Printout from Logistic Model: 'Agent Orange' Study

This printout gives identical results to those obtained in the preceding section, where these data were analysed using a direct approach. Note that the records in the data table for which the terms in the exposure column are 0 may be omitted, since their contribution to the likelihood does not involve the parameter b.

The method based on the logistic model has the advantage that it can handle additional covariates, in contrast to the direct method. Suppose that there is not just a single covariate designating exposure, but several covariates denoted by x_{i1}, x_{i2}, ... (for the cases) and y_{i1}, y_{i2}, ... (for the controls). Then it is easy to see, using the multiple logistic model given by Equation (5.6), that the likelihood for a matched case-control study given by Equation (8.5) generalises to

$$L = \prod_{i=1}^{n} \left(\frac{1}{1 + e^{-\Sigma b_j (x_{ij} - y_{ij})}} \right) \qquad (8.6)$$

Consequently the parameters b_j may be estimated by fitting the multiple logistic model, using a similar data layout as that for a single exposure variable.

To illustrate, consider the following matched case-control pairs selected from a study of risk factors for endometrial cancer among women living in a Los Angeles retirement village during the period 1971–75, cited by Breslow and Day (1980, page 167). Here the risk factor of interest is 'ever taken any oestrogen' and the case-control pairs are matched with respect to age, marital status, and date of entry to the retirement village. An additional covariate is available relating to gall bladder disease (present or absent), and the data layout is as follows:

Count	Case		Control		difference	
n	Est (x_1)	Gall (x_1)	Est (y_1)	Gall (y_2)	x_1-y_1	x_2-y_2
1	0	0	0	0	0	0
1	0	0	1	0	−1	0
1	0	0	1	1	−1	−1
21	1	0	0	0	1	0
18	1	0	1	0	0	0
1	1	0	0	1	1	−1
3	1	0	1	1	0	−1
2	0	1	0	0	0	1
1	0	1	1	0	−1	1
1	0	1	0	1	0	0
6	1	1	0	0	1	1
4	1	1	1	0	0	1
1	1	1	0	1	1	0
2	1	1	1	1	0	0

TABLE 8.13: Data for a Matched Case-control Study: Endometrial Cancer

The printouts from various logistic models fitted to these data are shown in Table 8.14. The first printout is obtained by including only oestrogen exposure, the second is obtained by omitting oestrogen exposure but including gall bladder disease status as a risk factor, while the third printout shows the model with both covariates included.

```
Logistic Regression Analyses
Response: Endometrial Cancer
  Col Name        Coeff  StErr p-value  Odds    95%      CI

   3 Oestrogen    2.269  0.606      0 9.666 2.945 31.728

df:62                  -2LogL:62.887                 #iter:6

  Col Name        Coeff  StErr p-value  Odds    95%      CI

   4 Gall         0.955  0.526  0.069 2.599  0.927 7.291

df:62                  -2LogL:83.654                 #iter:5

  Col Name        Coeff  StErr p-value  Odds    95%      CI

   3 Oestrogen    2.209  0.610  0.000 9.106 2.757 30.081
   4 Gall         0.695  0.616  0.259 2.003 0.599  6.695

df:61                  -2LogL:61.545                 #iter:6
```

TABLE 8.14: Matched Case-control Study ModelS: Endometrial Cancer

The bottom line of each printout contains the term $-2\log L$

(minus twice the log-likelihood), from which two models may be compared based on the difference, using the chi-squared distribution. Thus the model involving both oestrogen exposure and gall bladder disease as risk factors for endometrial cancer is not a great improvement on the model involving oestrogen exposure as the single risk factor (see also page 255 of Breslow and Day (1980)).

Covariates in pair-matched cohort studies may be handled in a similar way. Consider again the matched pairs cohort study investigating vasectomy as a risk factor for a myocardial infarct, reported by Walker (1982). In this study obesity and smoking status were also measured on each individual, as given in Table 8.15.

Pair	Which suffered MI	Vasectomised Obese	Vasectomised Smoker	Not Vasectomised Obese	Not Vasectomised Smoker
1	N	−	−	+	−
2	V	+	−	−	+
3	V	−	+	−	−
4	N	−	+	−	+
5	N	+	−	+	+
6	V	−	−	−	+
7	V	+	−	−	+
8	V	−	−	−	−
9	N	−	−	−	+
10	V	−	+	−	+
11	N	−	+	−	+
12	N	−	−	−	−
13	V	+	+	−	+
14	N	−	−	−	+
15	V	−	+	−	−
16	V	+	+	+	+
17	V	−	−	−	−
18	N	+	+	−	+
19	V	+	+	−	−
20	N	−	−	−	+
21	N	−	−	−	+
22	N	−	+	−	−
23	V	−	+	−	+
24	V	−	+	−	−
25	V	+	+	−	−
26	N	−	−	−	−
27	V	−	+	−	+
28	V	+	−	−	−
29	N	−	−	−	+
30	V	−	−	−	−
31	N	−	+	+	+
32	V	−	+	−	−
33	N	+	−	−	−
34	N	+	−	−	+
35	V	−	−	−	−
36	V	−	−	−	−

TABLE 8.15: 36 Data Pairs from Matched Cohort Study

This table shows which of the two men (vasectomised, V; or not vasectomised, N) had the infarct, with the symbols + and – denoting the presence/absence of obesity and smoking condition.

Data such as these may be analysed using a logistic model, based on the following theory. Suppose that the joint effects of an exposure E and a covariate F on an outcome D are given by the logistic model

$$Prob[D \mid E,F] = \frac{1}{1 + e^{-(a+bx+cy)}} \qquad (8.7)$$

where x is the value of the exposure of interest E, assumed to be dichotomous (so that x is 1 if the subject is exposed and 0 otherwise), and y is the value of a variable F (not necessarily dichotomous). Consider a pair of subjects, matched so that each member has the same value of F. If their outcomes are independent, the probability that the exposed member has a positive outcome and the nonexposed member has a negative outcome is

$$Prob[DE,\bar{D}\bar{E}] = \frac{1}{1 + e^{-(a+b+cy)}} \times \left(1 - \frac{1}{1 + e^{-(a+cy)}}\right)$$

Similarly the probability of the converse is

$$Prob[\bar{D}E,D\bar{E}] = \left(1 - \frac{1}{1 + e^{-(a+b+cy)}}\right) \times \frac{1}{1 + e^{-(a+cy)}}$$

As for a pair matched case-control study, it is reasonable to assume that all the relevant information is contained is these discordant outcomes. Given just these two possibilities, the probability that the exposed member gets the positive outcome is

$$\frac{Prob[DE,\bar{D}\bar{E}]}{Prob[DE,\bar{D}\bar{E}] + Prob[\bar{D}E,D\bar{E}]} = \frac{e^{-(a+cy)}}{e^{-(a+b+cy)} + e^{-(a+cy)}}$$

$$= \frac{1}{1 + e^{-b}}$$

Suppose that the outcome for pair i is denoted by d_i, which takes the value 1 if the exposed member experiences the event and 0 if the nonexposed member does. Then the likelihood of the data is

$$L = \prod_{i=1}^{n} \left(\frac{1}{1 + e^{-b}}\right)^{d_i} \times \left(\frac{e^{-b}}{1 + e^{-b}}\right)^{1-d_i} \qquad (8.8)$$

Now if measurements are taken on additional covariates having values x_{i1}, x_{i2}, ... (for the exposed members) and y_{i1}, y_{i2}, ... (for the nonexposed members) in pair i, the same argument may be used to show that the likelihood is

$$L = \prod_{i=1}^{n} \left(\frac{1}{1 + e^{-b - \Sigma b_j(x_{ij} - y_{ij})}} \right)^{d_i} \times \left(\frac{e^{-b - \Sigma b_j(x_{ij} - y_{ij})}}{1 + e^{-b - \Sigma b_j(x_{ij} - y_{ij})}} \right)^{1 - d_i} \qquad (8.9)$$

This likelihood is again a special case of the likelihood function based on the unconditional logistic regression model given by Equation (8.4), so you may compute the maximum likelihood estimate of the log odds ratio b and the other parameters by fitting a logistic regression to a data array containing the exposed – nonexposed covariate differences as predictors. The estimate of b is the constant term in this model.

The printout obtained as a result of fitting the logistic regression model to the data in Table 8.15 is shown in Table 8.16. The adjusted odds ratio for the effect of the vasectomy is not very different from the crude estimate based on the data given in Table 8.11. However the printout shows that smoking is a risk factor for myocardial infarct in this target population. While there is no evidence to suggest that obesity is a risk factor, the estimated confidence interval is very wide.

```
Logistic Regression Analysis
Response: Myocardial Infarct
  Col Name        Coeff StErr p-val  Odds   95%     CI
```

Col	Name	Coeff	StErr	p-val	Odds	95%	CI
0	Vasectomy	0.198	0.404	0.623	1.219	0.552	2.691
3	Obese	1.178	0.780	0.131	3.249	0.705	14.982
4	Smoker	1.421	0.638	0.026	4.142	1.187	14.456

```
df:33   Dev:41.285  %(0):44.444  #iter:8  RSq: 0.165
```

TABLE 8.16: Logistic Model Fitted to Matched Pairs Cohort Study

4: Before–After Studies

Matched data also arise from *before–after* studies. In this type of study measurements are taken on individuals both before and after some exposure (such as a treatment or intervention), so you have a pair of matched observations on each subject. The outcome might be smoking status and the intervention an anti-smoking campaign.

In a cohort study all the subjects have not experienced the outcome when they enter the study, so only their outcomes at the end of

the study are relevant. In a case-control study the cases have experienced the outcome and the controls have not, and their outcome
status in the past is not relevant. In contrast, a before–after study
looks at outcomes both before and after an exposure period. For such
a study to be informative, it must be possible for outcomes to move in
both directions.

As an illustration, consider the following hypothetical example
involving 100 subjects classified by smoking status before and after an
advertising campaign:

		Before Campaign	
		Smokers	Non-smokers
After	Smokers	9	22
	Non-smokers	30	39

The proportion of smokers is reduced from 39% to 31% because 30 smokers have given up smoking and only 22 have taken up
the habit. Using McNemar's test (which is appropriate for matched
data such as these) the chi-squared value is $(30 - 22)^2/(30 + 22) = 1.23$,
which is not statistically significant.

Now look at the data in the following table: the number of subjects is still 100, the proportion of smokers still changes from 39% to
31%, but McNemar's test now gives the highly significant statistic of
$(9 - 1)^2/(9 + 1)$, or 6.4.

		Before Campaign	
		Smokers	Non-smokers
After	Smokers	30	1
	Non-smokers	9	60

Thus some care is needed in the interpretation of data from
before-after studies. The result is affected to a large extent by the
amount of correlation in the outcomes before and after the exposure.
A further complication with analysing data such as these is that the
numbers in the diagonal affect the proportions but not the statistical
significance of the before-after comparison. For example, if the diagonal counts were 391 and 599 instead of 30 and 60 (giving a total
count of 1000 subjects rather than 100), the change in the proportion
of smokers would be minuscule (from 40% before to 39.2% after) but
the p-value given by McNemar's test would remain the same.

Usually there is a control group in a before–after study. If there is no such unexposed group, you cannot tell whether a change is due to the exposure in question or to some other factor. The presence of a control group allows you to isolate the effect of the exposure after adjusting for the correlation in outcomes within each subject.

As an illustration, consider a study undertaken by Dr Amornrath Podhipak in 1991 involving 281 pharmacists in Bangkok. The outcome was type of prescription (effective or ineffective) for a child suffering from diarrhoea. The exposure was presence or absence of a training program designed to improve the prescription habits of pharmacists. The data are as follows, where type of prescription is coded as 0 (ineffective) or 1 (effective):

	treated prescription before 0	1			control prescription before 0	1
after 0	92	20		after 0	80	11
after 1	23	32		after 1	9	14

These data may now be analysed using the methods described in Chapters 4 and 5. The outcome *after* the intervention is taken as a dichotomous response variable, the exposure to the treatment is the risk factor of interest, and the outcome *before* the intervention is a stratification variable. The data layout is shown in Table 8.17 with computer printout obtained by fitting the logistic regression model.

count	total	E	X
23	115	1	0
32	52	1	1
9	89	0	0
14	25	0	1

```
Logistic Regression Analysis
Response: ep-after
 Col Name        Coeff  StErr p-value  Odds   95%     CI

   0 Constant   -1.463  0.221  0.000
   3 treated     0.563  0.314  0.073 1.756 0.948  3.250
   4 ep-before   2.046  0.304  0.000 7.735 4.264 14.033

df:1   Dev:0.778   %(0):72.242   #iter:7      RSq: 0.986
```

TABLE 8.17: Logistic Model fitted to Before–after Study

The interpretation of this printout is straightforward. The treatment effect is not statistically significant, having a 95% confidence interval (0.95, 3.25) for the corresponding odds ratio. Furthermore the outcome status before treatment is a strong predictor of the outcome status after treatment (odds ratio 7.7), as you would expect.

5: 1:M Matched Case-Control Studies

All of the matched designs considered so far in this chapter have involved matched pairs in which one member of the pair has one exposure (or outcome, for a case-control design) and the other member has a different exposure (or outcome).

Various generalisations are possible. If you wish to analyse three or more different exposures or treatments (or outcomes) using a matched design, triplets or quadruplets (or n-tuplets in general) would be needed. However, the analysis of these more complex designs is a specialised topic and is not considered here. A more common extension of the pair-matched design involves matching one case or treated subject with several controls. Such designs are called *1:M matched designs*.

If the cases and controls in a case-control study are equally prevalent, the most efficient design involves matching each case with exactly one control. However, it often happens that controls are more prevalent than cases, and the efficiency of the study may then be increased, up to a point, by matching each case with several controls.

Consider a matched case-control study in which case i is matched to m_i controls. Suppose that case i has covariates x_{ij} and the matched controls have covariates y_{ikj}, $k = 1, 2, ..., m_i$, $(j = 1, 2, ...)$. The likelihood may be derived using a simple extension of the argument for the situation in which m_i is 1. The expression for this likelihood is

$$L = \prod_{i=1}^{n} \left(\frac{1}{1 + \sum_{k=1}^{m_i} e^{-\Sigma b_j(x_{ij} - y_{ikj})}} \right) \qquad (8.10)$$

This expression is not a special case of Equation (8.4), so it is not possible to estimate the coefficients in the model using logistic regression. However it turns out that Equation (8.10) is a special case of the likelihood corresponding to the stratified proportional hazards model, described in Chapter 7. The correspondence is as follows.

Suppose you have stratified survival data where each stratum has precisely one subject experiencing the event whose failure time is no greater than any of the other (censored) survival times in the stratum. These strata correspond to the matched sets in the case-control study, where the case corresponds to the subject in the stratum who experienced the event and the controls correspond to the subjects with censored survival times.

To analyse data from a 1:M matched case-control study, the data should be structured in the usual way with one record for each subject. In addition to the covariates, separate columns need to be included for (1) an indicator of case-control status (1 for a case and 0 for a control, say), (2) an index specifying the stratum, and (3) an artificial variable for the 'survival time' which simply could be a constant such as 1.

As an illustration, consider the complete set of data from the endometrial cancer study among women living in a retirement community near Los Angeles cited by Breslow and Day (1980, pages 290–296) and originally reported by Mack et al (1976). The cases all had endometrial cancer and the controls were age-matched (within 5 years) women at risk of the disease living in the community at the same time. The exposure of interest was oestrogen use, and covariates were gall bladder disease occurrence, hypertension occurrence, and non-oestrogen drug use. Following the prescription described above, the first ten data records may be structured as follows:

Case	Age	Gall	HyperT	Estrogen	Other	Stratum	'SurvT'
1	74	0	0	1	1	1	1
0	75	0	0	0	0	1	1
0	74	0	0	0	0	1	1
0	74	0	0	0	0	1	1
0	75	0	0	1	1	1	1
1	67	0	0	1	1	2	1
0	67	0	0	1	0	2	1
0	67	0	1	0	1	2	1
0	67	0	0	1	0	2	1
0	68	0	0	1	1	2	1
.	..	.	.	.	.	.	.

Note that the extra column labelled '$SurvT$' has been attached to allow for survival times in the stratified survival analysis model. The output shown in the top panel of Table 8.18 is now obtained after fitting the model. This shows that oestrogen use is a significant risk factor for the disease and the presence of gall bladder disease is a significant covariate, but the other two covariates do not contribute.

As expected in view of the fact that the data are matched with respect to age, the age effect is minimal, showing only the effect of the small age variation (less than 5 years) within the cells. The possibility exists of an interaction between the two risk factors, and this is confirmed by the printout in the bottom panel of Table 8.18.

```
Stratified Cox Regression Analyses on matched 5-tuples
Response: Endometrial cancer case
 Col Name         Coeff  StErr p-val HazRatio    95%     CI
```

Col	Name	Coeff	StErr	p-val	HazRatio	95%	CI
2	Age	-0.290	0.249	0.244	0.748	0.459	1.220
3	Gall	1.302	0.413	0.002	3.677	1.636	8.264
4	HyperT	-0.126	0.350	0.718	0.881	0.444	1.748
5	Estrogen	1.958	0.460	0.000	7.086	2.874	17.468
6	Other	0.745	0.514	0.147	2.106	0.770	5.765

```
  n:315        %Cen: 80.000      -2LogL:154.121         #iter:8
```

Col	Name	Coeff	StErr	p-val	HazRat	95%	CI
3	Gall	2.894	0.883	0.001	18.072	3.202	102.010
5	Estrogen	2.700	0.612	0.000	14.882	4.487	49.360
9	Gall.Estr	-2.053	0.995	0.039	0.128	0.018	0.902

```
  n:315        %Cen: 80.000      -2LogL:153.461         #iter:7
```

TABLE 8.18: Matched Analysis of Endometrial Cancer Study

This result may be compared with that obtained by simply fitting a logistic regression model to the data, using the same covariates, thus ignoring the matching. The corresponding printout is shown in Table 8.19. As you can see the results are very similar, suggesting that there is little if any correlation between the exposures within the matched pairs.

```
Logistic Regression Analysis
Response: Endometrial cancer case
 Col Name        Coeff  StErr  p-val   Odds   95%      CI
```

Col	Name	Coeff	StErr	p-val	Odds	95%	CI
0	Constant	-3.663	0.585	0.000			
3	Gall	2.970	0.847	0.000	19.497	3.709	102.480
5	Estrogen	2.715	0.612	0.000	15.106	4.555	50.092
9	Gall.Estr	-2.230	0.943	0.018	0.108	0.017	0.682

```
 df:311     Dev:265.628    %(0):80     #iter:10    RSq: 0.157
```

TABLE 8.19: Unmatched Analsyis of Endometrial Cancer Study

Recall that a subset of these data, which included all 63 cases and just 63 of the 252 controls, provided an illustration of the analysis of data from a matched pairs case-control study (Table 8.14). In that analysis oestrogen use, but not gall bladder disease, was found to be a risk factor, although the interaction effect was not examined. Given that we have found no advantage in doing the matched analysis, let us repeat the unmatched analysis leading to Table 8.19 on the subset. Table 8.20 shows the resulting printout.

```
Logistic Regression Analysis
Response: Endometrial cancer case
   Col Name         Coeff  StErr  p-val   Odds    95%      CI
```

Col	Name	Coeff	StErr	p-val	Odds	95%	CI
0	Constant	-2.303	0.606	0.000			
3	Gall	2.590	0.975	0.008	13.333	1.974	90.070
5	Estrogen	2.886	0.657	0.000	17.916	4.944	64.931
9	Gall.Estr	-2.400	1.122	0.032	0.091	0.010	0.818

```
df:122     Dev:140.785    %(0):50    #iter:10    RSq: 0.194
```

TABLE 8.20: Unmatched Analysis of Subset of Subjects

The estimates given in Table 8.20, based on just 63 control subjects, are not too different from those in Table 8.19 using the data from all 252 controls: the standard errors do not increase very much. This demonstrates that there is only a marginal gain in increasing the sample size in a case-control study when the number of cases is fixed.

6: Pros and Cons of Matching

Matching is the special case of stratification that arises when the strata are small and balanced. Pair matching is the most extreme situation, since the matched sets then contain exactly one subject of each type – one case and one control for a case-control study, or one exposed and one nonexposed subject for a cohort study. Pair matching is the most efficient study design you can have, since each matched pair contains the maximum amount of information obtainable concerning the association of interest based on the minimal number of subjects.

Matching can improve efficiency by reducing the sample size needed for a study. Consider again the Ille-et-Vilaine case-control study of risk factors for oesophageal cancer introduced in Section 5 of Chapter 4. In this study there were 200 cases and 775 controls, 975 subjects in all, and the subjects were classified into six 10-year age groups. The allocation of these cases and controls to the nine tobacco and alcohol exposure combinations is given in Table 4.14. Table 8.21 shows an allocation that results from a particular selection of 200 of the 775 controls in which the number of cases and controls is the same in each age group. The members of the subset are selected by reducing the number of controls in each exposure cell by approximately the same proportion within each age group. Thus, for example, the numbers of controls in the nine exposure classes corresponding to age group 35–44 are reduced by a factor of approximately 20.

		Tobacco: 0-9			Tobacco 10-19			Tobacco: 20+		
Alcohol:		0-39	40-79	80+	0-39	40-79	80+	0-39	40-79	80+
Age										
25-34	Case	0	0	0	0	0	1	0	0	0
	Cont.	1	0	0	0	0	0	0	0	0
35-44	Case	0	0	2	1	3	0	0	1	2
	Cont.	3	2	0	1	1	0	1	1	0
45-54	Case	1	6	7	0	4	9	0	10	9
	Cont.	12	8	4	5	5	3	4	3	2
55-64	Case	2	9	14	3	6	14	7	7	14
	Cont.	22	15	6	9	7	3	5	7	2
65-74	Case	5	17	9	4	3	5	2	5	5
	Cont.	22	9	4	5	4	5	4	2	0
75+	Case	1	2	3	2	1	2	1	1	0
	Cont.	7	1	0	2	1	0	1	1	0

TABLE 8.21: Reduced Data from Oesophageal Cancer Case-control Study

Fitting a logistic regression model to the reduced data set gives the printout in Table 8.22, which may be compared with Table 5.13.

```
Logistic Regression Analysis
Response: oesophageal cancer
  Col Name          Coeff  StErr p-value   Odds    95%      CI
```

Col	Name	Coeff	StErr	p-value	Odds	95%	CI
0	Constant	-1.458	1.825	0.424			
4	Tob 10-19	0.381	0.273	0.163	1.464	0.857	2.499
5	Tob 20+	1.070	0.293	0.000	2.917	1.644	5.176
7	Alc 40-79	1.438	0.280	0.000	4.212	2.431	7.298
8	Alc 80+	2.559	0.314	0.000	12.919	6.988	23.885
10	Age 35-44	-0.172	1.896	0.928	0.842	0.020	34.589
11	Age 45-54	-0.521	1.835	0.776	0.594	0.016	21.589
12	Age 55-64	-0.332	1.827	0.856	0.717	0.020	25.766
13	Age 65-74	0.029	1.831	0.987	1.030	0.028	37.240
14	Age 75+	0.327	1.870	0.861	1.387	0.035	54.214

```
df:35    Dev:37.734    %(0):50    #iter:12    RSq:0.726
```

TABLE 8.22: Fit of Logistic Model to Reduced Oesophageal Cancer Data

The odds ratios for the tobacco and alcohol exposures are much the same. The confidence intervals based on the smaller study are only a little wider, indicating that not much efficiency has been lost by reducing the number of controls from 775 to 200. The most noticeable difference is that the age effects have disappeared.

Matching with respect to a covariate means that it is not possible to assess the covariate as a risk factor, for the simple reason that the values of the covariate are the same for the two or more subjects in the same set. In epidemiological studies it is common to match with respect to age. While age is usually strongly associated with the outcome of interest, this association is rarely of interest in its own right. As you saw with the oesophageal cancer study, age matching can improve the efficiency by reducing the number of controls needed to measure an association of interest with a specified accuracy.

Another advantage of matching is that it can reduce confounding bias. This concept may be illustrated by the Berkeley graduate admission data shown in Table 4.7. Suppose that the sample size in this study is reduced by taking approximately the same numbers of men and women applicants from each department. This will involve selecting only one in eight of the men from Department A, one in 20 from Department B, and half of the women from departments C and E, giving the data as shown in the next table.

		Accepted		Rejected	
		Men	Women	Men	Women
	A	64	89	39	19
	B	18	17	10	8
Dept.	C	120	101	205	196
	D	138	131	279	244
	E	53	47	138	149
	F	22	24	351	317
Total		415	409	1022	933

Using the totals the crude odds ratio is $(415 \times 933)/(409 \times 1022)$, or 0.926, with a 95% confidence interval (0.79, 1.09). This is very close to the Mantel–Haenszel adjusted odds ratio (0.932, 95% CI: 0.78–1.11), so the confounding has been eliminated.

Matching can reduce confounding bias, but it does not necessarily eliminate it. Consider the following hypothetical example of a cohort study to investigate an association between an exposure of interest E and an outcome D with a dichotomous covariate F.

	F			$\bar{F}$			Aggregated	
	E	$\bar{E}$		E	$\bar{E}$		E	$\bar{E}$
D	91	61	D	32	7	D	123	68
$\bar{D}$	9	39	$\bar{D}$	68	93	$\bar{D}$	77	132

OR = 6.5   OR = 6.2   OR = 3.1

Note that F is a confounder, although the allocation to exposure at each level of the covariate is completely balanced. However, if the covariate is included in the analysis an unbiased estimate of the odds ratio is obtained. A ramification of this result is that randomisation in a clinical trial does not necessarily eliminate confounding: it is still desirable to include all relevant risk factors in the statistical analysis.

But what about risk factors that are difficult or impossible to measure? Take socio-economic status: this is thought to be a risk factor for many epidemiological outcomes, but it is difficult to measure numerically. Another example is the skill of a surgeon, which may affect the outcome of a transplant operation where the exposure of interest might be treatment with a new drug.

One approach is to match with respect to the suspected confounder. This is possible even when the confounder cannot be measured numerically. Thus the strata could contain, for example, individuals having the same socioeconomic status, or patients operated on by the same surgeon.

To summarise, matching on a covariate which is an independent risk factor can accomplish two objectives. First, it can improve the efficiency of the study by eliminating the variation in outcomes associated with the covariate. Second, if the covariate is also associated with the risk factor of interest (and is thus a confounder) matching on it can reduce (but not necessarily eliminate) the confounding effect of the covariate. If the covariate is measurable, its effect can be accounted for in the analysis. But if a covariate is not measurable both a matched design and matched analysis is necessary to cope with it.

Overmatching

The term 'overmatching' has been used to describe situations where matching is unnecessary or counterproductive. If a covariate is not an independent risk factor for the outcome of interest, matching on it is simply a waste of effort. If a covariate is not an independent risk factor for the outcome but is associated with the risk factor of interest, matching on it is both wasteful and inefficient, since it will tend to create noninformative matched sets whose members have the same exposure. (As an illustration of this situation, Michael et al (1984, pages 121–131) describe a hypothetical case-control study to investigate an association between chronic bronchitis and air pollution where the subjects were matched on social class by choosing each case-control pair to reside in the same census tract. However, a null result was found because subjects residing in the same census tract experienced very similar air pollution exposures.)

When a covariate is associated with both the risk factor and the outcome, matching with respect to it can introduce a bias. In the strict sense, the term *overmatching* is used to describe this situation.

As an illustration of overmatching, consider a case-control study investigating poor nutrition as a risk factor for molar pregnancy in Filipino women, and suppose that the cases and controls are matched with respect to their mid upper arm circumference (MUAC) measurements, a covariate found in previous studies to be associated with both the risk factor and the outcome. The following tables give the results obtained from a hypothetical study:

Low MUAC				*High MUAC*				*Aggregated*		
	Nutrition				Nutrition				Nutrition	
	poor	good			poor	good			poor	good
case	90	20		case	10	45		case	100	65
cont	45	10		cont	20	90		cont	65	100
	OR = 1				OR = 1				OR = 3.3	

The matched analysis adjusts for the confounding effect of mid upper arm circumference and finds no association between nutritional status and the outcome. Now suppose that a woman's mid upper arm circumference measurement is determined jointly by nutritional status and molar pregnancy, that is, it is really an outcome that occurs as a result of these risk factors, rather than a determinant of molar pregnancy. In this case the correct result is that nutritional status is associated with molar pregnancy and matching on the covariate will fail to discover this association.

The moral of this example is that it is dangerous to match without an adequate understanding of the subject matter.

Summary

This chapter has discussed matching, a design technique aimed at improving the efficiency of a study and reducing confounding.

Matched Pairs

The most extreme matched design involves matched pairs. In a case-control study each case is individually matched with a control subject as alike as possible to the case, and their exposure to the risk factor of interest is measured. Only pairs with discordant exposures contain information about the odds ratio, estimated by the ratio of positive to negative discordant pairs. If there is no correlation within matched pairs the matched and nonmatched analyses give similar results.

Logistic Modelling

The logistic model may be used to handle covariates in matched designs. In a pair matched case-control study the data may be analysed by fitting the model to unit responses with predictors defined as the case-control differences and the constant term omitted, and a similar kind of analysis may be used for pair matched cohort studies.

Before–After Studies

Before–after studies involve the comparison of measurements taken on subjects before and after some treatment or exposure of interest, and may be regarded as a special case of matching. In these studies a control group is needed to distinguish a placebo effect from a real effect.

1:M Matched Designs

The logistic model may also be used for matched designs with one case and several controls in each matched set, and the data may be analysed using a stratified proportional hazards model, but there is a limit to the gain in efficiency you get by increasing the number of controls per case.

Pros and Cons of Matching

Matching is wasteful if the matching factor is not associated with the outcome, and can be inefficient when the matching factor is associated with the risk factor of interest. It can even introduce bias if a matching covariate is an outcome rather than a risk factor.

Exercises

Exercise 8.1: Data were obtained from a cohort study involving 480 subjects designed to investigate an association between a dichotomous exposure (E) and a disease outcome (D) in a target population.

The study actually comprised 240 twins, and the following results were obtained (A = first-born twin, B = second-born):

		ED	$\overline{ED}$	$\overline{E}D$	$\overline{ED}$	
	ED	26	8	21	23	78
	$\overline{ED}$	13	13	3	13	42
A	$\overline{ED}$	23	7	14	13	57
	$\overline{ED}$	11	19	4	29	62
		73	47	42	78	

B

How would you go about analysing these data? Undertake any analyses that are relevant.

Suppose that you wish to undertake a matched case-control study based on these data, by selecting only those twins where one developed the outcome and the other did not. Carry out the (matched) analysis using logistic regression.

Now suppose you are told that the matching carries no information since the twins were created using a computer matching program which just grouped randomly.

(a) Carry out the unmatched analysis on the same data that you used for the matched analysis.

(b) Carry out the analysis on the complete set of data.

(c) Check for correlation within the matched pairs.

Exercise 8.2: Using the data in Table 8.13, fit a model containing an interaction between oestrogen use and gall bladder disease (in addition to the main effects) and compare your result with the printouts in (a) the bottom panel of Table 8.14 and (b) Table 8.20.

Exercise 8.3: A study was conducted on smoking behaviour in children from two counties in Ontario and reported by Kalbfleisch and Lawless (1985). Those in one county received educational material about smoking during a two month period while those in the other group did not. At the beginning of the period there were 98 children in the treatment group who had never smoked, 18 current smokers and nine who had once smoked but had quit, compared with 64, 16 and eight, respectively, in these categories in the control group. At the end of the two months five of the 98 never-smokers in the treatment group had taken up the habit, but two of these had then quit, ten of the 18 smokers had stopped, and only one of the nine quitters had relapsed. In the control group three of the never-smokers started smoking but two of these quit before the two months, eight of the 16 current smokers quit, and only one of the eight quitters had relapsed.

Is the treatment effective? Do an appropriate analysis.

References

Breslow, N.E. & N.E. Day (1980): *Statistical Methods in Cancer Research: Volume I – The Analysis of Case-Control Studies.* IARC. Lyon.

Donovan, J.W., R MacLennan & M. Adena (1984): Vietnam service and the risk of congenital anomalies; a case-control study, *Medical Journal of Australia* **140**, pp 394–397.

Kalbfleisch, J.D. & J.F. Lawless (1985): The analysis of panel data under a Markov assumption, *Journal of the American Statistical Association*, **80**, pages 863–871.

Mack, T.M., M.C. Pike, B.E. Henderson, R.I. Pfeffer, V.R. Gerkins, B.S. Arthur & S.E. Brown (1976): Estrogens and endometrial cancer in a retirement community, *New England Journal of Medicine*, **294**, pages 1262–1267.

McNemar, Q. (1947): Note on the sampling error of the difference between correlated proportions or percentages, *Psychometrica*, **12**, pages 153–157.

Michael, M, W.T. Boyce & A.J. Wilcox (1984): *Biomedical Bestiary: An Epidemiologic Guide to Flaws and Fallacies in the Medical Literature.* Little Brown and Company. Boston/Toronto.

Sleigh, A., R. Hoff, K. Mott, M. Bannetto, T. Maisk de Paiva, J. de Sousa Pedrosa & I. Sherlock (1982): Comparison of filtration staining (Bell) and thick smear (Kato) for the detection and quantitation of *Schistosoma mansoni* eggs in faeces, *Transactions of the Royal Society of Tropical Medicine and Hygiene*, **76**, pages 403–406.

Walker, A.M. (1982): Efficient assessment of confound effects in matched follow-up studies, *Applied Statistics*, **31**, pages 293–297.

9

SAMPLE SIZE

1: Introduction

Basic methods of analysis of epidemiological data are covered in the preceding chapters. These methods include the estimation and comparison of odds ratios using Mantel–Haenszel methods, modelling the risk of disease outcome using logistic and Poisson regression, survival analysis, and methods for improved design using stratification and matching.

In this chapter the problem of sample size determination is addressed. The choice of sample size for a study is one of the most important questions faced by the epidemiologist. It is just as important as the choice of study design, because it can influence this design choice.

The problem of choosing a sample size arises because of the statistical variation in repeated samples taken from a population. A sample can never exactly reflect the characteristics of a population. You may know from official statistics, for example, that in the most recent census 55% of 20–24 year old males in a region of a developing country have completed secondary school. But if you were to take a sample of, say, 100 men in this age group from various villages in the region, it is unlikely that you would find that exactly 55 men in the sample had completed secondary school. The number might be 50. Another sample would in all likelihood give a different number again. Even if your sample were carefully chosen to be unbiased (using some scientific sampling procedure, possibly involving stratification), there is no way you can guarantee that the proportion of men in the sample completing secondary school will be exactly 0.55. This is the nature of sampling variation: it cannot be eliminated.

Sampling error cannot be eliminated, but it can be reduced by choosing a sufficiently large sample. If your sample size were increased from 100 to 1000 randomly selected men aged 20–24 years, you would expect to find the proportion with a secondary school education to be closer to the population percentage (55% in this case). And if you took a very large sample of 10 000, say, you would expect its proportion to be quite close to the population value.

It is true that larger samples give more accurate estimates of population parameters, but there is a limit to the feasible sample size, and accuracy of estimation must be weighed against the cost of recruitment of subjects and data collection and analysis. If an estimate of a population characteristic is more accurate than it needs to be, it could mean that resources have been wasted answering a question with *too much* certainty when it would have been better to design the next study to investigate another, consequential, question.

In the following sections some formulas are given for computing sample sizes. These formulas tell how large a sample is needed to obtain (a) specified accuracy of an estimate of a population parameter, or (b) specified power of detection of a difference between estimates of parameters based on samples from different populations. Conversely, these formulas may also be used to gauge the accuracy and detection power obtainable from samples of specified size.

Before deriving the formulas, it is necessary to define some technical terms, including the precision of an estimate, types of errors, statistical power, and minimal worthwhile difference.

Sample size and power calculations are of particular importance when conducting clinical trials, and a special section of this chapter is devoted to such calculations. The chapter concludes with an introduction to the related concept of meta-analysis.

2: Precision of an Estimate

The precision of an estimate is related to its confidence interval. If the confidence interval is symmetric (as is usually the case when estimating a population mean), the precision is defined as half its width, and in this case the term *absolute precision* is used. If the parameter of interest is a ratio or rate, the precision may be defined as the percentage by which the lower bound of the (two-sided) confidence interval is less than the estimate, and the term *relative precision* is used.

The precision of an estimate thus depends on the probability content of the confidence interval (usually 95%). In general this probability content is represented as $1 - \alpha$, so that α is 0.05 for a 95% confidence interval.

The formula for the precision of an estimate depends only on $Z_{\alpha/2}$, the value on the horizontal axis of the normal distribution above which the area is $\alpha/2$, and the standard error of the estimate. This formula is based on the assumption that the sample size is large enough for the estimate to be normally distributed. For a symmetric confidence interval the derivation is as follows.

Denote the estimate by E and its standard error by SE. The confidence interval ranges from $E - Z_{\alpha/2}SE$ to $E + Z_{\alpha/2}SE$, so half its width is just $Z_{\alpha/2}SE$. Consequently the absolute precision is given by the expression

$$d = Z_{\alpha/2} SE \qquad (9.1)$$

This formula can be re-expressed to give the required sample size in terms of the absolute precision, provided the relationship between the standard error and the sample size is known. For example, if you are estimating a population mean, the standard error is given by the formula $SE = \sigma/\sqrt{n}$, where σ is the population standard deviation. (Note that this is the actual standard error based on knowledge of the population standard deviation, rather than its estimated value based on the sample standard deviation, which is given by Equation (2.4).) Thus Equation (9.1) gives (after straightforward algebraic manipulation)

$$n = Z_{\alpha/2}^2 \frac{\sigma^2}{d^2} \qquad (9.2)$$

Suppose you wish to estimate the mean age of menarche (first menstruation) of women in a certain target population, based on data obtained from a survey, and you would like your estimate to be within two months of the population mean, with 95% confidence. Based on data already collected from other populations, the standard deviation of ages at menarche in your population is expected to be 12 months. How large a sample do you need to take?

Equation (9.2) provides a solution to this problem. The absolute precision is $d = 2$ months, and the assumed population standard deviation is $\sigma = 12$ months. For 95% confidence, $Z_{\alpha/2} = 1.96$. Substituting these values, Equation (9.2) gives $n = (1.96)^2 12^2 / 2^2 = 138.3$, so you will need to survey 138 subjects. Note that this result is very sensitive to the standard deviation: if σ were 8 months instead of 12 months only 61 subjects would be needed to estimate the population mean with the same precision.

Now suppose you are interested in estimating the difference between two population means. Assuming that these populations have common standard deviation σ, the standard error of the difference in sample means is given by the exact version of Equation (2.5), that is

$$SE(\bar{y}_1 - \bar{y}_2) = \sigma \sqrt{\frac{1}{n_1} + \frac{1}{n_2}} \qquad (9.3)$$

where n_1 and n_2 are the sample sizes. Again using Equation (9.1) you can show that the required size of the first sample is

$$n_1 = \left(1 + \frac{1}{r}\right) Z_{\alpha/2}^2 \frac{\sigma^2}{d^2} \qquad (9.4)$$

where r is n_2/n_1, the ratio of the sizes of sample 2 to sample 1. Note that if sample 2 is very large (effectively a population) Equation (9.4) reduces to Equation (9.2), since only sample 1 is lacking in precision. Another important special case arises when the two samples are equal, when the size of each sample is twice that given by Equation (9.2).

As an illustration of the application of Equation (9.4), imagine that you wish to evaluate the effectiveness of a new drug aimed at alleviating pain among patients with diagnosed duodenal ulcers. The study is double blind and involves randomising patients to two groups, a treatment group and a placebo control group, and the responses of the patients in each group are recorded on a standardised scale with scores ranging from 0 to 5. You would like to be able to measure the difference in the mean scores in the treated and control populations with a precision of 0.25 with 95% confidence. Further, since the doctors recruiting the patients are reluctant to allocate their patients to an inactive treatment, the study design requires that the number of patients in the treated group be double the number in the control group. Previous studies indicate that the standard deviation in response scores is approximately 0.8.

Putting all of this information into the symbols used in Equation (9.4), we have $Z_{\alpha/2} = 1.96$, $d = 0.25$, $\sigma = 0.8$, and $r = 0.5$. Substituting these values, Equation (9.4) gives $n_1 = 1.5 \times (1.96)^2 (0.8)^2 /(0.25)^2 = 59.0$, so you will need to have 59 patients in the treated group and half as many, 30, in the control group.

Similar formulas are obtainable for estimating proportions. When estimating a one-sample population proportion π using a sample proportion p, the standard error is estimated from the exact version of Equation (2.1) in which the estimate p is replaced by the population value π, that is,

$$SE(p) = \sqrt{\frac{\pi(1 - \pi)}{n}} \qquad (9.5)$$

so Equation (9.1) gives

$$n = Z_{\alpha/2}^2 \frac{\pi(1 - \pi)}{d^2} \qquad (9.6)$$

As an illustration of Equation (9.6), suppose you wish to estimate the prevalence of acute respiratory tract infection, with a precision of 5% ($d = 0.05$), in a target population comprising children aged 2–4 years in a particular region of a developing country. Since an estimate of π is not available until the survey has been carried out, Equation (9.6) does not tell you what sample size is needed. However the formula may still be used to get a range of sample sizes corresponding to various assumptions for the value of π. Thus choosing $\pi = 0.1$, Equation (9.6) gives $n = (1.96)^2 \, 0.1(1-0.1)/(0.05)^2 = 138$. (As before, α is taken to be 0.05 corresponding to 95% confidence.) Similarly if π is 0.2, n is 246, while if the prevalence is 0.3 a sample size of 323 is obtained from the formula. The sample size needed takes its maximum value when the prevalence is 0.5, and in this case is 384.

For a fixed absolute precision d, you can see from Equation (9.6) that the required sample size increases as π increases from 0 to 0.5, and then decreases in the same way as the prevalence approaches 1. This means that if you had absolutely no idea about the prevalence (or if you wished to estimate several different prevalences from the same survey), the safest approach to take is to assume that π is 0.5.

The *relative precision* could be defined as the absolute precision (d) divided by the value of the parameter (θ, say). Using Equation (9.1), this gives the relative precision as

$$\varepsilon = Z_{\alpha/2} \frac{SE}{\theta} \qquad (9.7)$$

Consequently the relative precision of a proportion or prevalence is obtained by replacing d in Equation (9.6) by π (the population parameter) times ε to give

$$n = Z_{\alpha/2}^2 \frac{1 - \pi}{\varepsilon^2 \pi} \qquad (9.8)$$

Thus, in the preceding example, to estimate the prevalence of acute respiratory tract infection with a relative precision of 20% ($\varepsilon = 0.2$), the sample size required when the population prevalence is 0.1 is, from Equation (9.8), $n = (1.96)^2 \{(1-0.1)/0.1\}/(0.2)^2 = 864$. If π is 0.2, n is 384, when π is 0.3 n is 224, and when π is 0.5 the sample size needed is 96. Note that in this case the sample size needed decreases as the prevalence increases to 0.5, the reverse of the rule for the absolute precision of an estimate of prevalence.

The formula for the relative precision of an odds ratio in a case-control study may be obtained using the above arguments. First, note that a formula for the standard error of the (natural) logarithm of the odds ratio is (Equation (4.2))

$$SE(\ln OR) = \sqrt{\frac{1}{a} + \frac{1}{b} + \frac{1}{c} + \frac{1}{d}}$$

where a and b are the numbers of cases in the exposed and non-exposed groups, and c and d are the numbers of controls in these groups. Let $n_1 = a + b$ denote the total number of cases and $n_2 = c + d$ the total number of controls. The proportion of the cases exposed is thus $p_1 = a/(a + b)$, while $p_2 = c/(c + d)$ is the proportion of controls exposed. The standard error can thus be expressed in terms of p_1, p_2, n_1, and r (the ratio of controls to cases, n_2/n_1) as

$$SE(\ln OR) = \sqrt{\frac{1}{p_1(1-p_1)} + \frac{1}{rp_2(1-p_2)}} \times \frac{1}{\sqrt{n_1}} \qquad (9.9)$$

A confidence interval for an odds ratio is

$$OR\exp(-Z_{\alpha/2}SE), \quad OR\exp(Z_{\alpha/2}SE) \qquad (9.10)$$

Now if the relative precision is ε this interval is

$$(1-\varepsilon)OR, \quad (1-\varepsilon)^{-1}OR$$

and substituting the expression for the standard error from Equation (9.9), the required number of cases to give a specified relative precision is thus

$$n_1 = \left(\frac{Z_{\alpha/2}}{\ln(1-\varepsilon)}\right)^2 \left(\frac{1}{p_1(1-p_1)} + \frac{1}{rp_2(1-p_2)}\right) \qquad (9.11)$$

In practice you may not have direct information concerning both p_1 and p_2, but you may have information about the likely size of the odds ratio. If so the information about the odds ratio may be used to compute a corresponding value of the unknown proportion using Equation (4.1), which may be re-expressed as

$$OR = \frac{p_1(1-p_2)}{p_2(1-p_1)}$$

Thus p_1 is given by

$$p_1 = \frac{p_2}{p_2 + (1 - p_2)/OR} \qquad (9.12)$$

and the formula for p_2 in terms of p_1 is similar.

To illustrate the application of these formulas, suppose you wish to undertake a case-control study to determine the odds ratio, to a precision of 25% of its true value with 95% certainty, for the association between cholera and exposure to contaminated water in a refugee camp. You may assume that 30% of the residents are using water from the contaminated source and that the population odds ratio is at least 2. What sample sizes are needed in the cholera and control groups? (Assume equal numbers of cases and controls.)

The estimated proportion of exposed controls (p_2) is 0.3. Assuming the odds ratio is 2, Equation (9.12) gives $p_1 = 0.462$. Since the number of controls is equal to the number of cases, the allocation ratio r is 1. For 95% certainty $Z_{\alpha/2}$ is 1.96, and the relative precision (ε) is 0.25. Putting these values into Equation (9.11), the number of cases required is $n_1 = \{1.96 / \ln(0.75)\}^2 \{1/(0.462 \times 0.538) + 1/(0.3 \times 0.7)\}$, or 407.8. Thus 408 cases and 408 controls are needed.

Note that if you need to make an assumption about the value of the odds ratio to calculate an unknown proportion of exposed cases or controls, you should choose the value that is closest to 1 and is still consistent with what you expect. For example, taking the estimated odds ratio to be 1 in Equation (9.12) gives $p_1 = p_2 = 0.3$, and the estimate of n_1 is then 442. The lack of information about the odds ratio has increased the required number of cases from 408 to 442.

You can see that all of these sample size formulas are of the same form, namely that the required sample size is of the form

$$n = Z_{\alpha/2}^2 \frac{k}{\Delta^2} \qquad (9.13)$$

where $Z_{\alpha/2}$ is the appropriate normal deviate (1.96 if $\alpha = 0.05$), k is a constant that depends on the nature of the parameter being estimated, and Δ is a measure of the tolerance allowed (the greater the precision required, the smaller the tolerance). Thus the required sample size is inversely proportional to the square of the allowable tolerance.

3: Power of a Study

Comparative studies usually involve not just trying to get a confidence interval for some parameter of interest, but also testing a null hypothesis that two or more treatments or types of exposure are the same. In this case what is required is the minimum sample size needed to detect a *worthwhile* benefit.

The most crucial factor to consider when doing a sample size calculation for a comparative study is the minimum size of the benefit worth detecting. Take, for example, a disease such as a cancer where the usual response rate for the best currently available treatment is 20%. In this case a study undertaken to evaluate a proposed new treatment ought to find a benefit (that is, it ought to reject the null hypothesis) if its response rate were 30%. However, if the response rate for the new treatment were really only 21%, you would not be too concerned about failing to detect a benefit. In other words, an improvement in response rate from 20% to 30% is worth detecting, but an improvement from 20% to 21% is not. What benefit is worth detecting? The answer will also depend on the toxicity, cost, and inconvenience of the proposed treatment compared with the standard.

The worthwhile benefit must be distinguished from the *plausible* benefit, which is the benefit you think might be achieved based on experience (previous studies) and biological considerations. If the plausible benefit is less than the worthwhile benefit, there is little point in proceeding with the study. However, if the plausible benefit exceeds the worthwhile benefit, the sample size calculation should still be based on the worthwhile benefit, to ensure that optimistic expectations do not yield sample sizes too small to find important benefits. To illustrate this point, suppose that the new drug 'extramycin' is found to have a response rate of 30/40 (75%) for a cancer where the standard treatment only achieves a response rate of 20%. Now suppose that a Phase III (randomised control) trial is planned to compare extramycin with the standard treatment, and the investigators choose their sample size to have an 80% chance of detecting a difference of 55% (75% minus 20%), as this is considered plausible. If the actual difference were 30% there is a danger that this trial would give a null result, thus missing a clinically important benefit.

Another parameter that arises in comparative studies is the *statistical power* of the study. The power is the chance of detecting a worthwhile benefit which exists. In other words, the power is the chance of rejecting the null hypothesis when a worthwhile difference exists between the treatment groups. In the above hypothetical example, the power was stated to be 80% or 0.8. The power of a comparative study is usually denoted by $1 - \beta$, where β is the chance of failing to detect a difference, given that a worthwhile difference exists (in statistical terminology, the *Type II error*).

A formula may be derived relating the power of a study and the size of the worthwhile effect. The derivation is as follows. Suppose that the estimate of the effect is normally distributed with mean 0 and standard deviation SE_0 when the null hypothesis is true. Then the null hypothesis will be rejected whenever the magnitude of this estimate is greater than $Z_{\alpha/2}SE_0$, where α is the Type I error (usually 0.05 corresponding to $Z_{\alpha/2} = 1.96$).

Suppose now a worthwhile effect δ exists, in which case the distribution of the estimate is normal with mean δ and standard deviation SE_1. The distributions of the estimate for the two situations – (a) the null hypothesis is true, and (b) a worthwhile effect δ exists – are shown in the upper panel of Figure 9.1, whereas the lower panel shows the curve for situation (b) with the horizontal scale shifted so that the curve is centred at 0.

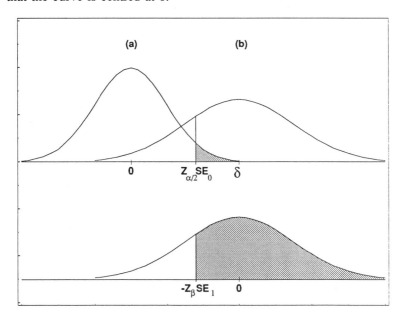

FIGURE 9.1: Distributions under Null and Alternatiive Hypotheses

From the picture in the top panel you can see that the distance between the minimum (positive) estimate needed to reject the null hypothesis ($Z_{\alpha/2}SE_0$) and the worthwhile effect (δ) is $\delta - Z_{\alpha/2} SE_0$. In the bottom panel you can see that power of the study, $1 - \beta$, is the (shaded) area to the right of the value $-Z_\beta SE_1$, and the distance from this point to the centre of the curve is thus $Z_\beta SE_1$. Since these two distances are the same, it follows that

$$\delta = SE_0 Z_{\alpha/2} + SE_1 Z_\beta \qquad (9.14)$$

Equation (9.14) is the basis for the calculation of sample sizes in comparative studies, and some special cases are now considered.

First, if two population means μ_1 and μ_2 are being compared, you would normally use the difference in the sample means as the basis for testing the null hypothesis. Whether or not the null hypothesis is true, this estimate has standard deviation given by Equation (9.3) (assuming a common standard deviation σ in the two populations), so Equation (9.14) becomes

$$\delta = (Z_{\alpha/2} + Z_\beta)\,\sigma \sqrt{\frac{1}{n_1} + \frac{1}{n_2}}$$

where δ is $\mu_2 - \mu_1$ (assuming $\mu_2 > \mu_1$). If the number of subjects in sample 2 is r times the number in sample 1 (that is, $n_2/n_1 = r$), the required size of sample 1 is thus

$$n_1 = \left(1 + \frac{1}{r}\right)(Z_{\alpha/2} + Z_\beta)^2 \frac{\sigma^2}{(\mu_2 - \mu_1)^2} \qquad (9.15)$$

Consider, for example, a randomised study to compare two treatments aimed at increasing weights of anorexic girls with a control treatment, where the minimal worthwhile benefit is an increase in mean weight of 5 pounds, and the standard deviation of weight changes is believed to be 7.5 pounds. Suppose that you wish to determine the sample sizes needed to achieve a power of 80% for each comparison. Assume an equal allocation to the three groups.

Given that there are three comparisons it is reasonable to take the type I error level to be 0.02 instead of the conventional level of 0.05. Thus α is 0.02 and β is 0.2, so $Z_{\alpha/2} = 2.33$ and $Z_\beta = 0.84$. Also σ is 7.5, $\mu_2 - \mu_1$ is 5, and r is 1. Substituting these values Equation (9.15) gives $n_1 = 2 \times (2.33 + 0.84)^2 (7.5)^2 / 5^2 = 45.2$. Conclude that 45 subjects are needed in each group, 135 in all.

Equation (9.14) also provides a formula for calculating the sample size needed to compare two proportions π_1 and π_2 with a given statistical power. In this case the standardised difference between the proportions in the two samples may be used as a test statistic. When the null hypothesis is true ($\pi_1 = \pi_2$), the estimated standard deviation of this difference is given by

$$SE_0 = \sqrt{p(1-p)} \sqrt{\frac{1}{n_1} + \frac{1}{n_2}} \qquad (9.16)$$

where p is the estimated proportion in the combined sample, that is

$$p = \frac{p_1 + rp_2}{1 + r} \quad (9.17)$$

On the other hand when π_1 is not equal to π_2, the formula for the estimated standard deviation is given by

$$SE_1 = \sqrt{\frac{p_1(1-p_1)}{n_1} + \frac{p_2(1-p_2)}{n_2}} \quad (9.18)$$

Substituting Equations (9.16) and (9.18) into (9.14), you get, after some algebraic manipulation,

$$n_1 = \frac{\left(Z_{\alpha/2} \sqrt{\left(1 + \frac{1}{r}\right)p(1-p)} + Z_\beta \sqrt{p_1(1-p_1) + \frac{p_2(1-p_2)}{r}} \right)^2}{(p_1 - p_2)^2} \quad (9.19)$$

where r is again the allocation ratio n_2/n_1.

Equation (9.19) is quite general: it applies to cross-sectional, cohort, and case-control studies. Some examples of its application are now given.

As a first illustration, consider a case-control study to compare the efficacy of a vaccine for the prevention of childhood tuberculosis with a placebo. Assume that 30% of the controls are not vaccinated. If the numbers of cases and controls are equal what sample size is needed to detect, with 80% power and 5% type I error, an odds ratio of at least 2 in the target population?

Note that the probability of exposure (that is, no vaccine given) for a control is $p_2 = 0.3$, and for an odds ratio of 2 Equation (9.12) gives $p_1 = 0.3/(0.3 + 0.7/2) = 0.462$. For an equal allocation of cases and controls r is 1 and p is $(0.3 + 0.462)/2 = 0.381$. Now α is 0.05 and β is 0.2, so $Z_{\alpha/2}$ and Z_β are 1.96 and 0.84 respectively, and Equation (9.19) gives $\{1.96\sqrt{(2 \times 0.381 \times 0.619)} + 0.84\sqrt{(0.462 \times 0.538 + 0.3 \times 0.7)}\}2$ divided by $(0.462 - 0.3)^2$ for n_1, which is 139.9. Conclude that 140 cases and 140 controls are required for the study.

As a second application of Equation (9.19), consider the randomised trial cited in Section 2 of Chapter 2, where 264 subjects who had had a mild myocardial infarct were randomised to hospital treatment or home care and 17 of those given home care died within six weeks compared with 14 in the hospital treated group. Assuming that a 20% reduction in the six-week mortality rate is a worthwhile benefit, how large a study would have been needed to gain 80% power?

Based on the study p is $31/264 = 0.117$, so assume p_1 is 0.106 and p_2 is 0.128, giving a relative risk close to 1.2. Equation (9.19) gives $\{1.96\sqrt{(2 \times 0.117 \times 0.873)} + 0.84\sqrt{(0.106 \times 0.894 + 0.128 \times 0.872)}\}^2$ divided by $(0.106 - 0.128)^2$, that is, 3346, for n_1, so a study size of almost 7000 is needed.

An alternative sample size formula for a case-control study may be derived using the natural logarithm of the odds ratio rather than the difference in proportions as the basis for the calculations. If OR_1 denotes the worthwhile odds ratio of interest, δ is $\ln(OR_1)$. The relevant standard error SE_1 is given by Equation (9.9), with the expression for SE_0 arising as the special case when $p_1 = p_2 = p$, where p is given by Equation (9.17). With these values, Equation (9.14) becomes

$$n_1 = \frac{\left(Z_{\alpha/2}\sqrt{\left(1 + \frac{1}{r}\right)\frac{1}{p(1-p)}} + Z_\beta\sqrt{\frac{1}{p_1(1-p_1)} + \frac{1}{rp_2(1-p_2)}} \right)^2}{(\ln OR)^2} \quad (9.20)$$

Equation (9.20) gives similar results to Equation (9.19). For the first example, where $OR_1 = 2$, $p_2 = 0.3$, $p_1 = 0.462$, $r = 1$, $p = 0.381$, $\alpha = 0.05$ and $\beta = 0.2$, it gives $n_1 = 140.5$, which is extremely close to the sample size calculated from Equation (9.19). For the second example Equation (9.20) gives $n_1 = 3337.3$, which is again very close to the result obtained from Equation (9.19).

If the study is a case-control study, the proportion p_2 could be used instead of the combined proportion p in Equation (9.19) or (9.20) on the grounds that the target population is likely to have many more controls than cases and consequently p_2 is a more accurate estimate in the absence of information about p_1. With this modification, the first example gives a reduced sample size requirement of 130 cases and 130 controls. However, a different result again is obtained from Equation (9.20), which gives $n_1 = n_2 = 152$ when p_2 is replaced by p. Clearly, further research is needed on the accuracy of these formulas.

The sample size equations given above all refer to unmatched studies. For matched designs similar formulas may be derived. First consider the comparison of two population means using a paired t-test (Equation (2.8)). The standard error is $\sigma_d/\sqrt{n_1}$, where n_1 ($= n_2$) is the number of matched pairs and σ_d is the standard deviation of the difference between an outcome in the treated or exposed group and the corresponding outcome in its matched control. Equation 9.14 gives

$$\delta = (Z_{\alpha/2} + Z_{\beta}) \frac{\sigma_d}{\sqrt{n_1}}$$

from which, using the fact that δ is $\mu_2 - \mu_1$, the formula

$$n_1 = (Z_{\alpha/2} + Z_{\beta})^2 \frac{\sigma_d^2}{(\mu_2 - \mu_1)^2} \tag{9.21}$$

is obtained.

It is instructive to express Equation (9.21) in terms of the correlation between the outcomes within the matched pair. If this correlation is ρ, the standard deviation of the difference in outcomes within a matched pair may be expressed as

$$\sigma_d^2 = 2\sigma^2 (1 - \rho)$$

where σ is the standard deviation of outcomes (assumed to be the same in each treatment population). Equation (9.21) then becomes

$$n_1 = 2(1 - \rho)(Z_{\alpha/2} + Z_{\beta})^2 \frac{\sigma^2}{(\mu_2 - \mu_1)^2} \tag{9.22}$$

It is interesting to compare Equation (9.22) with Equation (9.15), which applies to the analogous unmatched design. If the two groups are of equal size, r is 1, so the sample size requirement for a matched design can be less than or greater than that for the corresponding matched design, depending on whether the correlation is positive or negative. If the correlation is 0 it does not matter which design is used, at least for sample sizes large enough for the asymptotic theory underlying the formulas to apply. But if there is substantial correlation between outcomes within the matched pairs there is an appreciable reduction in the sample size needed to obtain a specified power. This fact explains why crossover designs are so efficient for comparing treatments in clinical trials.

The situation is rather different for matched studies involving dichotomous outcomes. In fact it turns out, as may be verified by substituting the formula for the standard error given by Equation (8.2) into Equation (9.14), that matched pair designs can lead to substantially larger sample sizes being required to achieve the same power as the corresponding unmatched design, even when there is very high positive correlation between outcomes (or between exposures for case-control studies). The reason for this apparent anomaly has been discussed in Chapter 8. In contrast to matched pairs of continuous measurements, matched pairs of dichotomous measurements with identical values provide very little, if any, information concerning the association of interest.

Note that in general the sample size needed for a comparative study is of the form

$$n = \frac{F}{\Delta^2} \qquad (9.23)$$

where Δ is a measure of the worthwhile effect and F is a constant which depends on the Type I and Type II errors α and β and the study design. You can see immediately from this formula that halving the size of the worthwhile effect means quadrupling the sample size.

Equations (9.15), (9.19), (9.20) and (9.22) give expressions for the sample size needed to achieve a specified power. It is also of interest to know how to calculate the power for a given sample size. To do this, you first turn the equation around to obtain an expression for Z_β in terms of the other parameters, and then use the inverse of the standardised normal probability function to get an expression for β.

As an illustration, consider Equation (9.15). The first step in this re-expression gives

$$Z_\beta = \frac{\mu_2 - \mu_1}{\sigma \sqrt{\dfrac{1}{n_1} + \dfrac{1}{n_2}}} - Z_{\alpha/2}$$

and, inverting the normal probability function, the second step gives the formula

$$\beta = 1 - \Phi \left(\frac{\mu_2 - \mu_1}{\sigma} \left(\frac{1}{n_1} + \frac{1}{n_2} \right)^{-\frac{1}{2}} - Z_{\alpha/2} \right) \qquad (9.24)$$

where $\Phi(z)$ is the cumulative normal distribution function.

As an illustration of the method, suppose that, in the investigation of weight gains of anorexic girls described in the discussion immediately following the derivation of Equation (9.15), you wish to calculate the power of a study designed to detect a weight difference of 5 pounds between two groups each containing 45 subjects, given a common standard deviation of 7.5 pounds for the weight decreases in each group, taking the Type I error to be 0.05. Substituting $n_1 = n_2 = 45$, $Z_{\alpha/2} = 1.96$, $\sigma = 7.5$, and $\mu_2 - \mu_1 = 5$, Equation (9.24) gives

$$\beta = \Phi \left(\frac{5}{7.5 \times \sqrt{2/45}} - 1.96 \right) = \Phi(1.20) = 0.885$$

so the power is 88.5%.

For the special case of equal allocation ($r = 1$) the formulas cited in the preceding sections are given, together with tables facilitating calculations, in Lwanga & Lemeshow (1991). Associated theory is also discussed in Lemeshow et al (1990).

4: Clinical Trials

Sample size calculations in epidemiological studies often lead to surprisingly large numbers of subjects being required. What if it is not feasible to recruit so many subjects in a reasonable time? Two options are available: (a) consider starting or joining a multi-centre or multinational study; (b) proceed with the study but plan on the need for a meta-analysis (described in Section 5 of this chapter) to eventually give a more definitive result, making sure that the report on the study contains all the relevant information needed for such an overview. It is also important to do a survey to discover how many eligible patients there are, and how many might be available for the proposed study.

Some special features that need to be considered when calculating sample sizes for clinical trials are now discussed. These features include the focus on comparing survival curves rather than proportions of outcomes, the effect of non-compliance, and relevant design issues.

Comparing Survival Curves

In a clinical trial involving the comparison of two or more treatments for a chronic disease, the number of subjects needed is usually so large that it will take many months or years to accrue these patients. The outcome event of interest (usually death or remission for degenerative diseases, or recovery for curable diseases) will usually not occur for several months or years later. Treatments are not usefully evaluated simply by comparing the proportions of events to occur in each treatment group, since these proportions will increase as the trial progresses and will eventually approach 1. For this reason the survival curves form the proper basis for comparison.

Provided the hazards are proportional, survival curves may be compared statistically using the logrank test, which involves comparing the observed numbers of events in each treatment group with expected numbers based on the assumption that the hazard functions are identical. The number of events occurring is simply obtained by multiplying the total sample size by an appropriate proportion. For this reason the sample size formula (9.19) for comparing two proportions may also be used in clinical trials whenever the proportional hazards assumption is met.

Rules of Thumb

Some rules of thumb are useful as a first step in a sample size calculation. One such rule is the *50:50 rule*, which states that to have an 80% chance of detecting a 50% relative reduction in the event rate, at least 50 events are needed in the control group. For example, if the usual (short-term) mortality were 10%, then 500 individuals will be needed in the control arm because this corresponds to $500 \times 0.1 = 50$ events. Note that if the event rate is low, large numbers of patients are required. Thus for a mortality rate of 2% in the control group with equal numbers in the two treatment groups, 5000 subjects need to be studied to detect a reduction to 1%.

The 50:50 rule follows directly from the sample size formula (Equation (9.19)) for comparing two proportions, simply by substituting $\alpha = 0.05$ (thus giving $Z_{\alpha/2} = 1.96$), $\beta = 0.2$ (that is, $Z_{\beta} = 0.84$), and $(p_1 - p_2)/p_1 = 0.5$ (so $p_2 = p_1/2$), to obtain an upper bound for the number of events in a treatment group $n_1 p_1$ as $6 (1.96 + 0.84)^2$, or 47. (This is actually an upper bound because it ignores factors in the numerator of the form $1 - p$).

For continuous outcomes such as blood pressure, the amount of variability in the measurements must be known. This variability may be expressed as a standard deviation, and represents the 'noise' through which you wish to detect the change: the greater the noise, the more difficult it is to detect a specified difference. A second rule of thumb applies only to continuous outcomes and is known as the *16 sigma over delta squared* ($16(\sigma/\delta)^2$) *rule*. It gives the sample size in each of two groups required to detect a difference in the means equal to δ, where σ is the standard deviation of the individual measurements in each group. As with the 50:50 rule, this rule assumes a type I error of $\alpha = 0.05$ and 80% power ($\beta = 0.2$). It is easily derived from the general sample size formula (9.15) for comparing the means of two normally distributed populations.

As an illustration of the $16(\sigma/\delta)^2$ rule, suppose you wish to detect a difference of 10 mm Hg in diastolic blood pressure between a treated group and a control group. Assuming that you know that the between-subject variability in diastolic blood pressure is 15 mm Hg, the sample size needed is $16(15/10)^2$, or 36 in each group, a total of 72 individuals altogether.

The Effect of Non-Compliance

The efficiency of a clinical trial is reduced substantially if subjects do not comply. To compensate for non-compliance, the sample size needs to be increased by a factor F, where

$$F = \frac{1}{(c_1 + c_2 - 1)^2} \qquad (9.25)$$

and c_1 and c_2 are the compliance rates (proportions of subjects receiving their allocated treatments).

As an illustration of the effect of non-compliance, consider a study conducted by the Multiple Risk Factor Intervention Trial (MRFIT) Group in 1977 which investigated smoking exposure for myocardial infarct outcome in men at elevated risk aged 35–57. Even though 12 886 subjects were followed up, only 10% of the control group and 26% of the intervention group gave up smoking, giving $F = 1/(0.9 + 0.26 - 1)^2$, or 39.06. Thus if compliance were 100%, to have the same efficiency the trial would have needed only 12886/29.06 = 330 subjects.

In trials requiring long term treatment, non-compliance can have an adverse effect on the required sample size, so designs are often considered which select a more compliant patient population. This can be accomplished by excluding patients considered to be unreliable or by having a run-in period prior to randomisation.

Design Issues

The design of a trial can affect the sample size required. Stratification and matching can reduce the sample size needed. By eliminating variation between subjects, a matched pairs design can reduce the required sample size substantially, particularly if the outcome is continuous rather than binary. This is why cross-over designs do not need large sample sizes.

Factorial designs can also be used effectively to reduce sample size requirements. With a 2-by-2 factorial design it is possible to answer two questions for not much more than the price of one. Sometimes it is desirable to compare three or more treatments instead of just two. A useful rule-of-thumb here is to work out the sample size requirement for the two-group comparison, and then use the same number of subjects per treatment arm in a multi-arm study as in a two-arm study. A more accurate rule is to reduce the Type I error rate, α, for each hypothesis to be tested, so that the overall error rate is still approximately α. This would mean, for example, that α should be approximately 0.02 for a three-arm study if you want an overall error rate of 0.05.

5: Meta-analysis

Although a team of investigators may begin their study with the aim
of settling a particular research question, in practice studies rarely
provide definitive answers. If a study has a large sample size and is
well designed it may provide an answer in itself to the question of
interest, but most studies simply provide evidence that contributes to
the body of scientific knowledge.

In practice epidemiological studies do not always have sufficient power to detect effects such as clinically important treatment differences. You saw in the preceding section an example where a randomised trial of 264 subjects was inconclusive, being nowhere near
large enough to detect a worthwhile difference in mortality between
two treatment groups. You also saw (in Chapter 8) that a case-control
study investigating a possible link between exposure to the 'agent
orange' defoliant by Vietnam soldiers and subsequent birth anomalies
in their children reported by Donovan et al (1984) involved over 8000
matched pairs yet still found an inconclusive 95% confidence interval
for the odds ratio (0.80–1.33).

It has been argued that studies should not be undertaken unless
they have sufficient power to be conclusive. Since large sample sizes
are often needed to achieve reasonable power, this criterion would rule
out small studies with inadequate power and force investigators to
undertake multi-centre studies.

In statistical language, failure to detect a true effect of interest
is called an error of the second kind or *Type II error*. There is also the
possibility that a study will find that an effect exists when it does not,
giving rise to an error of the first kind (*Type I error*). Although this
error is usually fixed at 5% by scientific convention and is unrelated
to the size of the study, it provides another argument against prolif-
erating smaller studies. The argument runs as follows. Suppose that a
large number, say 1000, studies are undertaken to investigate treatments for a specific disease, but in only 10% of them is the treatment
really effective. Suppose also that the studies in which the treatment
effect is real each have power 0.5 (corresponding to a Type II error of
50%). Then 50 of the 100 studies with the true effect will correctly
detect it. However, with a Type I error of 5%, 45 of the 900 studies in
which the treatment is ineffective will also report a (false) positive
result, giving only 50 correct positive conclusions out of 95 reported.
Note that increasing the power does not solve the problem here: even
if each study had 100% power the ratio of correct conclusions among
those reporting a positive result would rise to only 100 out of 145.

Small studies undertaken by individual investigators thus tend
to suffer from two problems, namely (a) the fact that a small study

may have insufficient power to detect a clinically important effect, and (b) the likelihood that many individual investigators are undertaking studies in which there is no effect present, yet 5% of them will still declare positive conclusions, most of which are false. These considerations have been used to justify expenditure of limited resources on large multi-centre studies rather than on small studies undertaken by individual investigators.

However, there are arguments on the other side. Large trials are costly and require a high degree of organisation. Should individual investigators really be discouraged from undertaking their own studies? Many patients may suffer and die needlessly from a disease despite the availability of a beneficial treatment, suggested as promising on the basis of several small but inconclusive trials, but withheld until its benefit is confirmed by a large definitive study.

There is a solution to the problem. It is possible to combine evidence from related studies and thus arrive at a conclusion. The method is called *meta-analysis*, and has been used extensively in social science research (see, for example, Hedges and Olkin (1985)).

Meta-analysis may be defined as the process of applying statistical methods to the problem of combining results from different analytic studies of the same research question. The term *overview* is also used to describe this process, particularly in the clinical trials context (see, for example, Peto (1987)). According to Peto, an overview involves an exhaustive process of obtaining individual results from clinical trials, whereas a meta-analysis could be a less exhaustive summary of results from the literature.

The first – and generally most difficult – step in meta-analysis is to decide which studies to include. Ideally each contributing study will involve investigating the same treatment or risk factor for the same disease on subjects drawn from the same target population, and each study will have the same type and quality of design. In practice all these factors vary, and you need to decide how much tolerance to allow in the inclusion criteria.

Biases can arise if all relevant studies are not included in a meta-analysis. *Publication bias* can arise if only published studies are included, because studies with negative findings are less likely to be published (see, for example, Easterbrook et al (1991)). Begg and Berlin (1988) advocated adoption by the scientific community of a 'policy agenda that will lead to the improved quality of published research data and so reduce the impact of bias'. They also suggested that the bias could be quantified more effectively if study registries were established, as advocated by Simes (1986, 1987). *Reviewer bias* can also arise if the overview team tends to select studies for the meta-analysis favouring a particular conclusion. Chalmers (1988) advocated that the evaluation of the quality and appropriateness of the studies included be performed blind to their outcomes.

As pointed out by Simes (1990), meta-analysis has weaknesses, namely (a) its ability to detect not only small effects but also small biases, (b) the difficulty of interpreting results from dissimilar studies (as discussed by Thompson and Pocock (1991), for example), and (c) the retrospective nature of the method. To avoid biases, Peto (1987) argues for restricting overviews to *randomised* trials. DerSimonian & Laird (1986) suggested using a statistical model incorporating random effects (generalising the usual fixed effects model) as a way of dealing with the heterogeneity, although Peto (1987) claims that the fixed effects models are more realistic. The retrospective nature of the method may be allowed for by doing meta-analyses *prospectively*, that is, by planning to undertake meta-analyses with research questions, inclusion criteria, outcomes, study designs and other relevant factors decided before the results of the individual studies are known.

For studies with binary outcomes, a simple statistical method for meta-analysis uses the Mantel–Haenszel approach for combining odds ratios described in Chapter 4. As an illustration, consider 16 studies of antihypertensive drug treatment for coronary heart disease (CHD) assembled by Collins et al (1990) and also considered from a cost-effectiveness viewpoint by Simes and Glasziou (1992), where the proportions of CHD events in each group are listed in Table 9.1. The table also contains the odds ratios for the individual studies. (The odds ratios are not defined for the two studies with no events in either the treated or control group.)

There are several methods for computing a combined odds ratio from data such as these. These include (1) application of the general, asymptotically valid, formulas given by Equations (4.4) and (4.5) to the individual log odds ratios, (2) the Mantel–Haenszel method (Equation (4.6), together with its variance estimate given by Equation (4.7)) for combining odds ratios, (3) logistic regression using the studies as strata, and (4) a method suggested by Richard Peto (outlined in Yusuf et al (1985)) which sums differences of observed and estimated numbers of events in the treated groups. The first three methods are described in Chapters 4 and 5.

Peto's method for combining data from separate studies may be described as follows. In a trial of N patients, where n are treated and d experience the outcome event, let O be the observed number of outcomes in the treated group. Assuming no treatment benefit, the expected number of outcomes in the treated group is $E = n \times d/N$, and the quantity $O - E$ would differ only randomly from 0, with variance $V = E(1 - n/N)(N - d)/(N - 1)$. (For example, using the data from the HDFP-1 study given in Table 9.1, $O - E$ is -21.98 and V is 100.94.)

| Study | Treated | | Control | | Odds Ratio |
	CHD	Total	CHD	Total	
VA-NHLBI	8	508	5	504	1.60
HDFP-1	191	3903	236	3922	0.80
OSLO	14	406	10	379	1.32
ANDPS	33	1721	33	1706	0.99
MRC	222	8700	234	8654	0.94
VA-2	11	186	13	194	0.88
USPHS	15	193	18	196	0.83
HDFP-2	61	1048	63	1004	0.92
HSCSG	7	233	12	219	0.53
VA-1	0	68	2	63	-
Wolff	0	45	0	42	-
Barraclough	1	58	2	58	0.49
Carter	2	49	2	48	0.98
HDFP-3	23	534	44	529	0.50
EWPHE	48	416	59	424	0.81
Coope	35	419	38	465	1.02

TABLE 9.1: CHD Incidences in Cardiovascular Studies

Assuming independent studies, an overall measure is obtained by adding up all the $O - E$ contributions, and the variance of this sum is the sum of the individual variances. A 95% confidence interval for the overall odds ratio is now given by

$$\exp\left(\Sigma(O_i - E_i)/\Sigma V_i \pm 1.96/\sqrt{\Sigma V_i}\right) \qquad (9.26)$$

The four methods are asymptotically equivalent. Table 9.2 shows a comparison based on the data in Table 9.1, and you can see that the results are very similar.

Method	Odds Ratio	95% CI
Asymptotic	0.868	0.780 - 0.966
Mantel–Haenszel	0.865	0.777 - 0.961
Logistic Model	0.862	0.775 - 0.957
Peto	0.865	0.778 - 0.961

TABLE 9.2: Comparison of Meta-analysis methods: CHD Data

Figure 9.2 graphs the odds ratios and 95% confidence intervals for the individual studies together with the meta-analysis result (where the confidence interval is represented by a diamond-shaped figure). The areas of the squares representing the values of the odds ratios increase with the weights given to the various studies; these are inversely proportional to the squares of the standard errors of the log odds ratios. Other things being equal, the weight increases linearly with the sample size of a study.

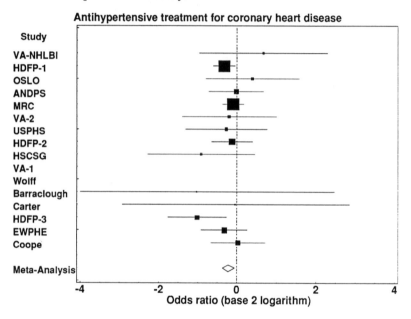

FIGURE 9.2: Odds Ratios for 16 Studies with Meta-analysis

The question of heterogeneity may be approached using the method for comparing odds ratios discussed in Chapter 4. It is also useful to graph residuals, obtained by standardising each log odds ratio by subtracting the overall log odds ratio and dividing by the individual standard errors. The plot of these residuals against normal scores is given in Figure 9.3. The superimposed line has slope 1, corresponding to homogeneity. You can thus see from Figure 9.3 that the studies contributing to the meta-analysis do not provide evidence of heterogeneity.

The method of combining outcomes based on Equations (4.4) and (4.5) is more general than the other methods since it may be applied to any measures of outcome, not just event counts. In survival analysis the outcome of interest for comparing two or more risk factors is a hazard ratio, and the meta-analysis may be performed using the associated beta-coefficients and standard errors from a model such as Cox's proportional hazards model. Note that these coefficients are usually adjusted for covariates. Some authors including Peto (1987) have argued that the *unadjusted* estimates should be preferred on practical grounds, particularly in view of the fact that in randomised studies the effect of adjusting for covariates is usually small. In Yusuf et al (1985) Peto has outlined another intuitively simple method for combining results from studies where the outcome is time to failure, by adding up the $O - E$ (observed minus expected) components to the logrank statistic from each study.

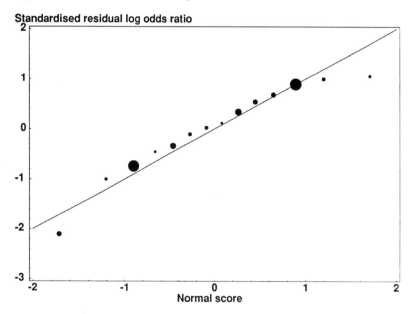

FIGURE 9.3: Standardised Residuals vs Normal Scores: CHD Studies

It is possible to adjust a meta-analysis to allow for covariates, such as dose level of a drug, which may vary between studies. As well, you may wish to allow for studies of varying quality, by carrying out a weighted regression analysis, with weight close to 0 for a very poor study and weight 1 for a perfectly good study.

A relatively recent innovation is *cumulative* meta-analysis, in which updated meta-analyses are successively carried out as each new study becomes available (see Lau et al (1992) and Antman et al (1992) for some discussion on this method).

Summary

In this chapter we have been concerned with quantifying statistical variability. This variability is due to the fact that studies have limited sample size, and is measured by the precision of population parameter estimates and the power of hypothesis tests.

Precision of an Estimate

The precision of an estimate of a population parameter may be expressed in either absolute or relative terms. The absolute precision is defined as half the width of a 95% confidence interval, while the relative precision is the percentage decrease in the parameter needed to reach the lower limit of its 95% confidence interval. To double the precision (that is, to reduce the width of the confidence interval by a factor of two) it is necessary to increase the size of the sample by a factor or four.

Power of a Study

The power of a study is associated with a null hypothesis, and is defined as the probability of rejecting the null hypothesis when a worthwhile effect really exists. The power depends on various factors including the sample size, the magnitude of the worthwhile effect, the variability in the measurements in the target population, the study design, and the type I error. An important consideration when planning a study is to choose a sufficient sample size to achieve a reasonable statistical power, and formulas are available.

Clinical Trials

Sample size determination is an important part of the protocol when planning a clinical trial, and special considerations arise due to censoring of survival data and non-compliance. Two rules of thumb are given.

Meta-Analysis

Meta-analysis is a method for combining results from different analytic studies of the same research question. The method can be used cumulatively and prospectively, and can enable a definitive conclusion to be reached from several inconclusive studies, provided inherent problems (reviewer bias, reporting bias, lack of homogeneity) have been addressed.

Exercises

Exercise 9.1: How large a sample is needed in order to estimate (a) a prevalence with an absolute precision of 2%, (b) a mean diastolic blood pressure with an absolute precision of 5 mm Hg (assuming the standard deviation of the distribution is 10 mm Hg)?

Exercise 9.2: A case-control study involves 100 cases and 200 controls, and the expected odds ratio based on previous studies is 1.5. Assuming 20% of cases are exposed, what is the expected relative precision?

Exercise 9.3: How many heart disease deaths need to be observed in the course of a randomised trial comparing two treatments if you wish to detect a decrease from 10% to 8% in heart disease mortality? Assume the type I error is 0.05.

Exercise 9.4: A crossover trial is proposed with 16 subjects to compare meditation therapy with a bronchodilator spray for increasing peak flow in subjects with asthma. Assuming the worthwhile effect is 10 ml/sec and the within-subject variability is 15 ml/sec, what is the power of the study, assuming the type I error is 0.02?

Exercise 9.5: A pharmaceutical company wishes to undertake a multi-centre study to compare their treatment for chronic duodenal ulcers with that of a competitor, and they wish to include a placebo treatment as well. However, some centres will only participate in the study if the placebo treatment is not available to their patients. Assuming that the response rate for an active treatment is 60% and that the worthwhile difference in response rate between active treatments is 10%, how many subjects need to receive an active treatment? (Assume a balanced allocation to active treatments, 80% power, and 5% type I error rate.)

  Now suppose that the expected placebo response rate is 40% and you wish to detect a 20% active treatment benefit over the placebo with 90% power, as well as the 10% treatment difference with 80% power, keeping the overall type I error rate to 5%. How many patients need to be given the placebo?

References

Antman, E.M., J. Lau, B. Kupelnick, F. Mosteller & T.C. Chalmers (1992): A comparison of results of meta-analyses of randomized control trials and recommendations of clinical experts,

JAMA, **268**(2), pages 240–248.

Begg, C.B. & J.A. Berlin (1988): Publication bias: a problem in interpreting medical data, *Journal of the Royal Statistical Society (A)* **151**, pages 419–445.

Chalmers, T.C. (1988): Meta-analysis. In Chalmers, T.C. (Ed) *Data Analysis for Clinical Medicine*. International University Press: Rome.

Collins, R., R. Peto, S. McMahon, P. Hebert, N.H. Fiebach, K.A. Eberlein, J.O. Taylor, J. Godwin, N. Qizibash & C.H. Hennekens (1990): Blood pressure, stroke, and coronary heart disease, Part 2, short-term reductions in blood pressure: overview of randomized trials in their epidemiological context, *The Lancet* **335**, pages 827–838.

DerSimonian, R. & N. Laird (1986): Meta-analysis in clinical trials, *Controlled Clinical Trials* **7**, pages 177–188.

Donovan, J.W., R MacLennan & M. Adena (1984): Vietnam service and the risk of congenital anomalies; a case-control study, *Medical Journal of Australia* **140**, pages 394–397.

Easterbrook, P.J., J.A. Berlin, R. Gopalan & D.R. Matthews (1991): Publication bias in clinical research, *The Lancet* **337** (8746), pages 867–872.

Hedges, L.V. & I. Olkin (1985): *Statistical Methods for Meta-analysis*. Academic Press. New York.

Lau, J, E.M. Antman, J. Jimenez-Silva, B. Kupelnick, F. Mosteller & T.C. Chalmers (1992): Cumulative meta-analysis of therapeutic trials for myocardial infarction, *The New England Journal of Medicine*, **327**(4), pages 248–254.

Lemeshow, S., D.W. Hosmer, J. Klar & S.K. Lwanga (1990): *Adequacy of sample size in health studies*. John Wiley & Sons. Chichester.

Lwanga, S.K. & S. Lemeshow (1991): *Sample size determination in health studies: A practical manual*. WHO. Geneva.

Peto, R. (1987): Why do we need systematic overviews of randomized trials? *Statistics in Medicine* **6**, pages 233–240.

Simes, R.J. (1986): Publication bias: the case for an international registry of clinical trials, *Journal of Clinical Oncology* **4**(10), pages 1529–1541.

Simes, R.J. (1987): Confronting publication bias: a cohort design for meta-analysis, *Statistics in Medicine* **6**, pages 11–29.

Simes, R.J. (1990): Meta-analysis: its importance in cost-effectiveness studies, *The Medical Journal of Australia* **153**, Supplement, pages S13–S16.

Simes, R.J. & P.P. Glasziou (1992): Meta-analysis and quality of evidence in the economic evaluation of drug trials, *Pharmo-Economics* **1**(4), pages 282–292.

Thompson, S.G. & S.J. Pocock (1991): Can meta-analyses be trusted? *The Lancet* **338**, pages 1127–1130.

Yusuf, S., R. Peto, J. Lewis, R. Collins & T. Sleight (1985): Beta blockade during and after myocardial infarction: an overview of the randomized trials, *Progress in Cardiovascular disease* **27**, pages 335–371.

APPENDIX

The following data arose from a case control study investigating risk factors for perinatal mortality among women in Ho Chi Minh City in 1992 undertaken by Dr Hieu in partial fulfilment of the requirements for his Master of Science degree in Epidemiology in the Faculty of Medicine at Prince of Songkla University in Southern Thailand.

The variables are ID (1–412), outcome (0 = lived, 1 = died), birth weight in grams, and mother's age and number of years of education.

1	1	1300	27	6		33	1	2400	24	6		65	1	3450	31	2
2	1	1500	34	3		34	1	4500	25	7		66	1	1000	34	6
3	1	1400	31	1		35	1	1900	23	2		67	1	1000	34	6
4	1	3400	19	6		36	1	1500	28	3		68	1	2500	44	2
5	1	2100	31	6		37	1	3000	27	2		69	1	1520	27	10
6	1	1200	28	1		38	1	2800	24	3		70	1	1100	28	10
7	1	2800	36	6		39	1	2300	24	6		71	1	4100	30	3
8	1	2200	26	10		40	1	2800	26	1		72	1	3000	27	6
9	1	3200	20	7		41	1	1200	31	6		73	1	3100	32	0
10	1	1100	24	4		42	1	3400	27	6		74	0	3100	21	7
11	1	3200	25	6		43	1	3000	20	10		75	0	2480	30	16
12	1	3500	41	1		44	1	1600	24	16		76	0	1920	27	6
13	1	3200	37	2		45	1	4100	32	10		77	0	1800	17	6
14	1	2200	42	6		46	1	1740	38	7		78	0	2700	30	10
15	1	1700	27	3		47	1	1500	27	1		79	0	2820	24	11
16	1	1400	38	0		48	1	1800	23	1		80	0	2820	36	8
17	1	1800	26	10		49	1	1470	25	6		81	0	3200	33	10
18	1	3100	32	4		50	1	1200	35	2		82	0	2850	38	6
19	1	1100	24	3		51	1	3660	30	10		83	0	3300	26	6
20	1	1200	23	1		52	1	3120	20	6		84	0	3240	22	6
21	1	2750	22	3		53	1	2250	36	6		85	0	3800	38	2
22	1	1500	22	13		54	1	2150	30	11		86	0	2900	29	7
23	1	3300	27	10		55	1	3300	29	5		87	0	3000	38	10
24	1	2100	23	2		56	1	4100	33	6		88	0	2790	34	6
25	1	1230	27	3		57	1	2900	31	10		89	0	2760	27	6
26	1	2610	26	0		58	1	1200	26	10		90	0	3700	22	3
27	1	1200	23	6		59	1	2500	19	6		91	0	2900	20	6
28	1	1200	22	2		60	1	2900	34	3		92	0	3600	20	7
29	1	1350	25	8		61	1	2200	40	1		93	0	2800	26	6
30	1	1230	24	6		62	1	1500	24	1		94	0	3200	33	6
31	1	1700	30	6		63	1	3700	22	6		95	0	3500	27	6
32	1	1100	22	4		64	1	1600	31	1		96	0	3100	32	6

97	0	3700	37	7	151	0	4100	32	6	205	0	3100	21	6
98	0	3600	38	6	152	0	3400	31	3	206	0	3000	38	5
99	0	3200	20	8	153	0	3600	34	1	207	0	2800	27	6
100	0	2950	27	10	154	0	3200	20	4	208	0	2800	27	6
101	0	3200	25	6	155	0	3000	23	6	209	0	3300	22	12
102	0	3500	31	3	156	0	3600	26	7	210	0	2900	34	6
103	0	2500	29	1	157	0	2700	23	6	211	0	3300	36	7
104	0	2800	31	6	158	0	4300	40	6	212	0	3800	28	10
105	0	2400	25	7	159	0	2700	25	8	213	0	2700	28	6
106	0	2800	25	6	160	0	2700	28	14	214	0	3000	25	6
107	0	3500	29	6	161	0	3150	36	10	215	0	3200	24	12
108	0	3180	28	9	162	0	2910	26	6	216	0	2900	30	10
109	0	2700	43	9	163	0	2940	25	6	217	0	2600	32	7
110	0	2800	21	10	164	0	3120	29	6	218	0	3000	19	3
111	0	3200	28	8	165	0	1950	19	6	219	0	2700	23	0
112	0	4200	25	6	166	0	3350	34	6	220	0	3400	24	6
113	0	3700	18	6	167	0	2300	30	16	221	0	3200	23	4
114	0	3500	31	6	168	0	2100	32	6	222	0	3540	32	6
115	0	3000	28	7	169	0	3390	37	7	223	0	1900	25	6
116	0	3100	24	6	170	0	3600	33	0	224	0	3200	36	7
117	0	2850	21	6	171	0	2600	20	11	225	0	2800	26	2
118	0	3350	33	10	172	0	3000	23	10	226	0	2900	27	13
119	0	3400	32	3	173	0	3100	27	10	227	0	3150	23	6
120	0	3000	33	1	174	0	2190	29	5	228	0	3100	21	10
121	0	3200	26	6	175	0	3700	26	16	229	0	3100	29	10
122	0	3000	28	3	176	0	2130	27	6	230	0	3100	27	5
123	0	3200	24	2	177	0	3000	25	10	231	0	3200	24	6
124	0	3100	25	6	178	0	2300	25	10	232	0	3200	25	2
125	0	3100	33	6	179	0	2700	32	10	233	0	1200	19	6
126	0	2900	24	7	180	0	3030	23	10	234	0	2800	26	7
127	0	2000	27	11	181	0	2900	35	6	235	0	3650	26	5
128	0	3300	29	8	182	0	2400	19	7	236	0	3450	35	10
129	0	3500	29	7	183	0	3500	30	7	237	0	3200	23	0
130	0	3400	24	7	184	0	2850	34	6	238	0	3800	26	6
131	0	2500	21	6	185	0	2100	25	11	239	0	3600	24	7
132	0	3200	32	6	186	0	2820	29	10	240	0	4000	26	2
133	0	1350	29	6	187	0	2900	24	9	241	0	2800	35	2
134	0	2900	32	8	188	0	3700	36	6	242	0	3800	30	2
135	0	1350	19	8	189	0	2830	37	7	243	0	2550	27	6
136	0	3100	43	6	190	0	2800	19	11	244	0	2900	32	5
137	0	3500	24	8	191	0	1911	31	6	245	0	3000	28	6
138	0	3200	18	6	192	0	2940	31	10	246	0	2730	29	10
139	0	2600	25	6	193	0	2070	37	6	247	0	2200	31	6
140	0	2900	30	2	194	0	3100	41	3	248	0	2900	30	3
141	0	3340	21	10	195	0	3100	41	10	249	0	3350	31	7
142	0	2700	25	4	196	0	2500	29	10	250	0	3000	30	6
143	0	2500	26	3	197	0	3900	29	11	251	0	2300	23	5
144	0	3200	34	3	198	0	3500	42	1	252	1	3000	23	2
145	0	3600	34	6	199	0	2300	35	5	253	0	2350	33	6
146	0	3400	38	0	200	0	3400	33	10	254	0	3200	19	6
147	0	3200	27	6	201	0	3700	25	7	255	0	2900	24	3
148	0	3400	25	6	202	0	2950	33	8	256	0	3000	20	6
149	0	3200	28	4	203	0	2900	33	6	257	0	2900	33	6
150	0	3500	32	6	204	0	3200	25	6	258	0	3000	24	12

259	0	2500	19	6		313	1	2700	27	4		367	0	3100	34	8
260	0	2350	23	10		314	1	1100	35	4		368	0	3150	25	6
261	0	2940	34	8		315	1	3500	32	1		369	0	3200	29	5
262	0	3540	32	6		316	1	1440	36	2		370	0	3000	20	7
263	0	2940	26	16		317	1	2040	30	6		371	0	3000	28	11
264	0	3000	20	3		318	1	2340	21	0		372	0	3300	37	2
265	0	3050	27	10		319	1	1400	28	0		373	0	2800	28	6
266	0	4300	24	10		320	1	1440	34	0		374	0	3350	42	5
267	0	2970	25	7		321	1	3180	28	7		375	0	2500	24	9
268	0	2900	33	10		322	1	3250	29	4		376	0	4200	29	7
269	0	1500	22	11		323	1	1200	22	5		377	0	3850	34	6
270	0	3100	24	10		324	1	3840	31	7		378	0	3000	27	5
271	0	2900	28	3		325	1	1400	23	2		379	0	2900	36	2
272	0	2700	37	0		326	0	2900	27	2		380	0	3510	32	8
273	0	3700	31	7		327	0	3650	35	6		381	0	3060	23	3
274	0	2200	19	0		328	0	3000	21	5		382	0	3150	30	10
275	0	2700	23	1		329	0	3000	30	7		383	0	3600	22	8
276	0	3250	35	5		330	0	2200	34	6		384	0	3250	28	12
277	0	2700	33	6		331	0	2150	18	5		385	0	3400	30	8
278	0	3300	35	1		332	0	3000	26	7		386	0	3180	38	3
279	0	2240	29	8		333	0	2900	25	8		387	0	2400	26	5
280	0	3270	37	1		334	0	2400	41	5		388	0	2220	24	6
281	0	3090	39	7		335	0	2300	22	6		389	0	2700	35	5
282	0	3100	25	7		336	0	3500	30	9		390	0	3060	30	6
283	0	2500	25	7		337	0	3100	32	5		391	0	3000	27	2
284	0	3350	24	6		338	0	2900	25	6		392	0	3600	22	5
285	0	2800	32	6		339	0	2300	29	7		393	0	3300	31	5
286	0	2850	19	6		340	0	2900	23	4		394	0	2500	21	6
287	0	2200	18	6		341	0	3800	33	5		395	0	2700	26	5
288	0	3300	23	10		342	0	3200	32	5		396	0	3500	24	9
289	0	3350	28	3		343	0	2600	21	10		397	0	2650	37	5
290	0	2300	37	3		344	0	3400	24	5		398	0	2700	30	2
291	0	3750	40	2		345	0	2400	37	12		399	0	3600	25	5
292	0	2760	35	4		346	0	3200	24	4		400	0	3100	27	5
293	0	3200	37	5		347	0	3500	26	12		401	0	3050	25	10
294	0	2490	26	6		348	0	2880	44	5		402	0	3050	32	0
295	0	2400	22	4		349	0	2820	32	4		403	0	2800	22	4
296	0	3000	33	4		350	0	2900	30	7		404	0	3300	27	8
297	1	2900	28	3		351	0	2700	35	4		405	0	2950	36	3
298	1	2300	28	5		352	0	3200	26	7		406	0	3250	31	16
299	1	1500	26	6		353	0	2450	23	8		407	0	2250	40	8
300	1	2700	24	4		354	0	3000	19	7		408	0	3300	23	3
301	1	3000	26	2		355	0	3000	28	8		409	0	3000	24	10
302	1	1000	27	4		356	0	2820	22	4		410	0	3000	24	7
303	1	1000	27	4		357	0	3000	32	2		411	0	4200	27	7
304	1	1550	26	3		358	0	2640	25	5		412	0	2500	28	3
305	1	1500	36	1		359	0	3300	40	2						
306	1	2700	31	0		360	0	3000	29	1						
307	1	1380	33	4		361	0	2700	21	4						
308	1	1250	31	7		362	0	3000	19	5						
309	1	1000	21	2		363	0	2300	25	3						
310	1	1380	23	3		364	0	2800	23	6						
311	1	1800	20	5		365	0	3000	34	0						
312	1	1800	36	5		366	0	3360	34	5						

INDEX

abortion 27
absolute precision 266
accidents to ships 175
accrual period 196
acute respiratory tract infection
  269
adenocarcinoma 219
adjusted odds ratio 105
adjusted survival curve 218, 225
age-adjusted mortality 14
age-adjusted odds ratios 118, 122
age effect 256, 259
age group 146, 165, 171
age of menarche 267
agent orange study 242, 247, 282
Agresti, A. 185, 187
AIDS 3
air pollution 261
albuminuria 23
alcohol consumption 117, 146,
  258
Altman, D.G. 2, 121
analysis of variance
  one-way 66
  by regression 82
  two-way 71
Anderson, S. 229
Anionwu, E. 56
anorexic girls 46, 57
anova plot 75–77
  of odds ratios 119
antigen mismatches 231
antihypertensive drug 284
Antman, E.M. 287
ANZ study 39
Armitage, P. 2, 15
Armstrong, K.L. 27
assumptions of regression 81, 87
association 4, 96
  between risk factors 103

asthma 289
Auer rods 216
Australian babies 242

bacteriuria 121
balance 10, 167
Bangkok pharmacists 253
Bartlett, M.S. 122
base 2 logarithm 40, 119
base 10 logarithm 40
baseline hazard function 214
baseline survival curve 214, 218
beach pollution 177
before–after study 47, 52, 251
Begg, C.B. 283
behavioural determinants 3
Bell method 244
Berkeley admissions 107, 141,
  259
Berry, D.A. 92
Berry, G. 2
bias 11, 167
binary data 95
birth anomalies 242, 282
birth defects 242
birth weight 9, 163, 190
Bishop, Y. 99, 160
bladder cancer deaths 112, 171
blinding 11
blood-letting 1
blood pressure 58
body fat 91
body weight 46, 57
box plot 43, 47, 53, 67, 68,
  72–73, 80, 88
breast cancer 10, 12, 39, 166
breast development 90
Breslow, N.E. 7, 8, 112, 117,
  159, 171, 247, 249, 255
British doctors 1, 113, 171

bronchitis 261
Bross, I.J.D. 11
Brown, B.W. 15
Buring, J.E. 122
burns to skin 229
Burns, K.C. 121
Byzantine coins 82

Cameron, E. 66
cancer, gastric 230
cancer research 7
cancer survival times 66
captopril 57
cardiovascular disease 28, 285
case-by-case data 160, 177
case-control study 8, 98
cases 8
categorical data 3
CD4 count 3
censored data 195-196
censored linear regression 221
censored status 200, 209
Chalmers, T.C. 283
chance error 11
checking assumptions 81, 87
chemotherapy 230
chi-squared test
    for association 38, 62, 64, 240
    for comparing models 151
    goodness-of-fit 142–143, 185
    homogeneity 106, 114
    logrank test 203
cholera 1, 6, 271
cholesterol level 97
chronic disease 279
cigarette smoking 122
clinic, as a risk factor 115
clinical importance 23–25
clinical trial 1, 9, 279
CMFP 39
cohort study 7, 23, 98
coin mintings 82
Collett, D. 155, 227
Collins, R. 284
combining results 63–64, 69
common odds ratio model 142
community study 9

comparing survival curves 200
confidence interval 15, 18–20,
    33–35
    for odds ratios 137
confounding 11, 13, 125, 260
    asymmetry of 134
    conditions for 103
    effect modification 102, 125,
        135
    examples 100–102, 117, 132
    inflation 101
    in 2-by-2 tables 99
    masking 101
    modelling 130
constant hazard 212
contaminated water 1, 271
contingency table 16, 62
continuous data 3
continuous outcomes 41
continuous predictors 162
control group 48
controls 8
controlled study 6, 9
coronary heart disease 97, 113,
    121, 284
correlated predictors 167
correlation 5, 14, 23
    between determinants 103
    in matched pairs 242, 244, 256,
        277
covariate 4
    stratifying by 223
Cox, D.R. 57, 155, 209
credibility 11
cricket data 155-156
Crilley, J. 174
crossover design 15, 281
crossover study 15, 237
cross-sectional survey 6
crude mortality 14
cumulative incidence ratio 110
cumulative meta-analysis 287
cumulative risk of failure 214

data layout
    case-by-case 160–161
    for linear regression 86

for logistic regression 130, 136,
  139–141, 147
for matched study 248
for Poisson regression 171–172
for survival analysis 188, 206,
  210
data transformation 51
data type 31
Dale, G. 52
Dawber, T.R. 8
Day, N.E. 7, 8, 112, 117, 159,
  171, 247, 249, 255
decreasing hazard 214
degrees of freedom 84, 142, 174
demographic factor 3
demographic research 14
departures, from psychiatric ward
  185–185
depression, 155
DerSimonian, R. 284
determinant 3
descriptive study 6
design issues in sample size 281
deviance 128, 131, 142, 162,
  164, 172
diarrhoea 3, 253
diastolic blood pressure 58, 289
dichotomous data 2, 95
difference in proportions 35
difference in risks 35
differential selection bias 12
Dinse, G.E. 229
discordant pairs 239
disease level 4, 181
Dobson, A.J. 155
Doll, R. 1, 12, 113, 171
Donovan, J.W. 242, 282
dopamine activity 41
double blind study 11
Downs syndrome 8
drug-taking, psychotropic 165
duodenal ulcers 10, 268, 289

ear infections 177
Easterbrook, P.J. 283
ectopic pregnancy 13

effect modification 102, 119,
  125, 135, 173
egg counts 51
eggshell thickness 78
electrode resistances 92
eligibility criteria 11
endometrial cancer 247, 255
environmental exposure 3
epidemiology 1
epileptics 167
ethical question 33
Evans, D.A. 121
Everett, B. 46
examples of confounding
  100–102, 117, 132
exercise 121
expected counts 38, 39, 203
experimental study 6, 9
exposure 16, 96
external validity 12
extra-marital coitus attitude 157
extra-Poisson variation 180
eye colour 64
Ezdinli, E. 56

factorial design 281
failure risk 212
failure times 197
false positive result 282
father–son pairs 183
F-statistic 67, 73
Feigl, P. 216
field trial 9, 10
Feinstein, A.R, 12
Fienberg, S.E., 157
Fisher, R.A. 10
Fligner, M.A. 71
follow-up 111
follow-up period 196
Framingham study 7
Freedman, D. 107
Freireich, E.J. 200
fun runners 52

gall bladder disease 247
gastric cancer 230
gastric freezing 10

Gavaskar, S. 155
gender 141, 165
genetic factor 3
GHQ score 165
Gilchrist, W. 191
glomerular filtration rate 23
Goodman, L.A. 64
goodness-of-fit 142–143, 177, 184
graft survival times 231
graphical assessment 215
graphical display 36, 42, 44, 45, 63, 64, 65, 69
graphing odds ratios 41, 63, 64, 65, 102, 109

haemoglobin level 56
haemophilia 3
Haenszel, W. 105
hair colour 64
Hampton, J.R. 32
Hand, D.J. 46, 185
Harrell, F. 212, 225
hazard 212
hazard function 208
  baseline 214
hazard ratio 209
health promotion 9
heart disease 7, 12, 55, 62
Hedges, L.V. 283
Henderson, R. 231
Hendy, M.F. 82
hepatitis 205, 215, 226
hierarchical model 150
Hieu, Dr 163
Hill, A.B. 1, 12, 113, 171
Hill, J.D. 32
Hills, M. 15
histogram 17, 52
HIV infection 3
Hoenig, J. 185
Holt, J.D. 229
home vs hospital care 32, 55, 275
homogeneity test 106
hookworm 51
horse kick deaths 192
Hosmer, D.W. 155

hypertension 57, 255
hypothesis testing 37

immunisation 235
incidence 4
incidence density 111, 170
incidence density ratio 112, 171, 201
inconclusive result 23, 243, 282
independent outcomes 180, 187, 202
independent risk factor 103, 117
independent samples 48
indicator variable 85–86, 141, 147
information bias 11
insect trapping experiment 191
instantaneous failure risk 212
integrated hazard function 212
interaction 136, 144–145, 149–150, 173
intercept in regression 80
interchangeable outcomes 181
internal validity 12
interquartile range 43
intervening event 196
intervening variable 4
intervention 9

jogging 7

Kalbfleisch, J. 203, 210, 219, 263
Kaplan, E.L. 198
Kaplan–Meier curve 198, 210–211, 219
  log transformation of 215
Kato–Katz method 244
Katzenstein, M. 9
Kirk, A.P. 205
Kimber, A.C. 229
Kleinbaum, D.G. 11, 97, 155, 192

Lau, J. 287
LBW cricket decisions 155–156
least squares line 80
Lemeshow, S. 279

length of stay in ward 185–186
leukemia 200, 209–210, 215,
  216, 221, 229
likelihood 209, 245, 250–251,
  254
linear hazard 212–214
linear regression 77, 80, 85
  censored 221
linear trend 114, 174, 206
logarithm transformation 17, 40,
  51, 63, 68, 104, 119
  of cell counts 52, 216
  of survival curve 215
logistic function 126
logistic modelling 245
logistic regression 2, 129,
  125–155, 159–190
logrank test 202–203, 279
longitudinal 7
loss to follow-up 196
Louis, P.C.A. 1
lung cancer 12, 219–220, 227
Lwanga, S.K. 279
lymphomas 56, 229

Maag, J.W. 155
McCullagh, P. 174, 185
McGregor, G.A. 57
Mack, T.M. 255
McNeil, D. 51
McNemar, Q. 240
McNemar's test 240, 243, 252
Makuch, R.W. 218
malformations 9
mammography 166
Mantel, N. 105
Mantel–Haenszel
  estimate of odds ratio 105, 118,
    202
  estimate of relative risk 113
  method 2, 95–120
    for meta-analysis 284–285
  test statistics 97, 106, 113, 203
mastectomy 10
matched analysis 240, 260
matched case-control study
  245, 254

matched cohort study 243
matched design 3, 260
  sample size for 277
matched pairs 15, 237, 238
matched studies 46
matching 15, 235–262
  pros and cons 257
maternal age 238
maximum likelihood estimator
  246
Mazess, R.B. 91
measurement 3
measures of association 96
median 43
median survival duration 199,
  219
medication 4
Meier, P. 11
Mellin, G.W. 9
menarche 267
menstrual regulation 12
meta-analysis 3, 279, 282–287
  comparison of methods 285
  graph of results 286
Miao, L.L. 10
Michael, M. 261
mid upper arm circumference
  261
mitozantrone 39
modelling covariates 85, 129, 216
modelling risk 164
model survival curves 214, 218
molar pregnancy 261
Morant, G.M. 167
mortality 1, 190
motion sickness 121
motor accident fatalities 112
MRFIT study 281
multi-arm study 281
multicategorical determinant 85
multi-centre study 279
multiple causes 4
multiple comparison 63, 74, 89
multiple determinants 7
multiple odds ratios 118,
  149–150
multiple outcomes 7, 39

multiple regression 85
multiple risk factors 115, 135, 146, 149–150
multivariate data 169
Murray, J.D. 165
myelogenous leukemia 216
myocardial infarct 32, 122, 244, 275

natural logarithm 40
Nelder, J.A. 174, 185
neonatal mortality 99, 115, 127, 130, 138, 160
Neyzi, O. 90
nitrous oxide 27
non-compliance 280
non-Hodgkins lymphoma 229
non-matched analysis 240
non-responders 12
normal distribution 17, 18, 22, 45
normal scores 45, 69, 152, 176, 223
normality assumption 46, 69, 223
Norton, P.G. 62
null hypothesis 5, 20, 37
number at risk 197

observational study 6
occupational exposure 3
odds ratio 5, 16, 35, 96–98
  adjusted 105, 116–118
  advantage of 97
  combined 104
  confidence interval 36, 138, 149
  crude vs adjusted estimate 103
  definition 16, 97
  from logistic model 127, 129
  from matched pairs 239
  multiple 118, 149
  standard error 35, 97, 138, 149
oesophageal cancer 117, 146, 257
oestrogen 247
oral contraceptives 121, 122
outcome 3, 16
outcome severity 181
outlier 43, 82, 151, 177, 180
overdispersion 180

overmatching 260
overview 279, 283

pain alleviation 268
pair matching 242, 257
paired data 46
paired t-test 48
parallel line plot 50, 53
parity 238
Pauling, L. 66
Payne, A.C. 183
PCB concentration 78
Pearson's chi-squared test 38, 62, 64
pelican eggs 78
performance status 219, 222
perinatal mortality 4, 163
person-times 111, 187, 201
Peto, R. 203, 283, 284, 287
pharmacists 253
phase I trial 10
phase II trial 10
phase III trial 10
phase IV trial 10
placebo 10, 11
plasma beta endorphin 52
plausible benefit 272
plum root cuttings 122
pneumonia 1
Pocock, S. 56, 205
Podhipak, Amornrath 253
Poisson distribution 182
Poisson regression 2, 159, 170
  in survival analysis 203
  with case-by-case data 166
poliomyelitis 11
Pollard, A.H. 14
pollution in beaches 177
poor nutrition 261
population parameter 1
postmenopausal hormone use 121
power 23, 271, 282
precision 15, 266
predicted probabilities 142
predicted values 81, 173
prednisolone therapy 205
Preece, D.A. 192

pregnant women 27, 156
pre-marital contraceptive use 157
prenatal care 4, 99, 115, 127, 130, 138, 160
Prentice, R.L. 203, 210, 219
pre-term birth 238
prevalence 4, 6, 166
preventative measure 3
printout
  logistic regression 128, 131, 142, 161
  Poisson regression 171, 172, 179, 204
probability 5
progabide drug 167
proportional hazards
  assumption 212, 215
  graphical assessment 215
  model 208–209, 214
  p-value for test 212
  stratified model 224
prospective 8
  meta-analysis 284
psychiatric patients 186
psychotics 41
psychotropic drug-taking 165
publication bias 283
p-value 15, 20

quadratic term in model 163
quartiles 43, 218

radiotherapy 230
random blood alcohol testing 112
randomisation 10, 33, 167, 260
randomised trial 32, 284
ranking null hypotheses 6
rare disease 8
rare risk factor 9
recall bias 11
recoding outcomes 181–182
reduced data 258, 259
reduced model 131, 148, 217
referent category 40, 62, 65, 116, 117, 144, 162, 177, 186
registry, of studies 283
regression, linear 77, 80, 85

coefficients 85
  one-way anova by 82
  model assumptions 81
regression to the mean 51
Reiss, I.L. 157
relative precision 266, 269
relative risk 5, 96, 110, 112
remission times 200
renal transplant 231
representative sample 11
reserpine 12
residual sum of squares 84
residuals 40, 45, 67, 221
  standardised 151, 176
resistances of electrodes 92
respiratory tract infection 269
response 2
retirement village women 247
reviewer bias 283
right-censored data 195
Risebrough, R.W. 78
risk 5, 96
risk difference 5, 96
risk factor 7
risk factors, independent 117, 133
risk modelling 164
risk, number at 197
risk rate 208
Rithsmithchai, Ms Skulrat 238
Robins, J.M. 105
Robins formula 105, 113
Rossing, P. 23
Rowland, A.S. 27
r-squared 84, 128, 131
rules of thumb 280

Sackett, D.L. 12
Salk vaccine 10
sample 6, 167
sample size 3, 19, 265–288
  rules of thumb 280
sampling 16
sampling variability 11, 15, 265
saturated model 146, 173
scatter plot 50, 79, 82, 84
schizophrenia 41
Schlesselman, J. 8

Schoenfeld, D. 212
Scottish children 64
Scotto, J. 192
screening 3, 9, 166
selection bias 11, 167
selection of subjects 9
Senn, S. 15
sensitivity 23, 244
severity of disease 181
sexual discrimination 107, 141
Shapiro, S. 122
Shapiro–Wilk test 46, 221
shell thickness 78
ship accidents 175
sickle cell disease 56
sickness in schoolchildren 236
silver content of coins 82
Simes, R.J. 283, 284
simulation 16
single blind study 11
SIRDS 190
skewness 46, 51, 216
skill of surgeon 260
skin cancer 192
skin grafts 229, 237
skull measurements 169
sleep patterns 27
Sleigh, A. 244
slope of regression line 80
smoking 7, 12, 113, 117, 122,
    157, 171, 249, 252, 263
snoring 62
Snow, J. 1, 6
social class 183, 261
socioeconomic status 90, 260
specificity 23
Stablein, D.M. 230
Stampfer, M.A. 121
standard deviation 17, 69, 74
standard error 18, 19, 34, 42
  of binomial proportion 143
  of difference in means 43
  of difference in proportions 35
  of incidence density ratio 112
  of log odds ratio 18, 239
  of log relative risk 110
  of proportion 34, 143

of regression coefficient 81
of sample mean 42
of weighted average 104
standardised residuals 151, 176,
    180, 287
standing, work activity 238
statistical assumptions 75
statistical methods 31–55, 61–92
statistical power 3, 23, 272
statistical significance 21, 23–25
Statistics 2
Sternberg, D.E. 41
stool samples 51
Strachan, D.P. 28
straight line 80
stratification 100, 108, 115, 125,
    140, 187, 201, 206, 223, 225,
    254
streptomycin 1
study types 6
sums of squares 67, 73, 84, 172
survival 212
survival analysis 3, 189, 195–228
  by logistic regression 188
survival curve 195, 197, 214
  adjusted 218, 225
  baseline 214, 218
  calculation 198
  comparison 200, 279
  model 214, 218
survival data 185
swimmers 71, 177
systematic error 11
systolic blood pressure 58

tail area of distribution 20, 22
target population 6, 9, 11, 16,
    159, 167, 283
Thai mothers 239
thalidomide 9
Thall, P.F. 167
Thompson, S.G. 284
Tibetan skulls 169
tied failure times 200
time-stratified PH model 225
tobacco consumption 117, 146,
    258

toxicity 39
transformation
   of data 51, 222
   of odds ratio 17
treatment 4
tuberculosis 1, 275
Tukey, J.W. 43
Turkish girls 90
Tuyns, A.J. 117
twins 236
two-by-two table 32, 37–38, 79, 95
   of matched pairs 238, 243
two sample *t*-test 42–44
two standard error rule 34
type I error 23, 282
type II error 23, 272, 282

unadjusted estimates 287
United Nations 14

vaccine 10, 275
validity 12
Van Vliet, P.K. 190

variability 11
vasectomy 244
Vietnam veterans 242, 282
virginity 157
vomit data 121

Walker, A.M. 244, 249
Water Board 177
weight gain 46, 57
weighted average 104
Weinberg, C.R. 156
white blood cells 216
white blood counts 216, 218, 229
Winer, B.J. 75
withdrawal from study 196
Woolson, R.F. 229
work activity 238
worthwhile benefit 272
worthwhile effect 23–25

Yusuf, S. 284

z-score 21–22, 108, 143, 172